Sound Studio Construction on a Budget

F. Alton Everest

McGraw-Hill

New York San Francisco Washington, D.C. Auckland Bogotá Caracas Lisbon London
Madrid Mexico City Milan Montreal New Delhi San Juan Singapore Sydney Tokyo Toronto

Library of Congress Cataloging-in-Publication Data
Everest, F. Alton (Frederick Alton), date.
    Sound studio construction on a budget / F. Alton Everest.
        p.   cm.
    Includes index.
    ISBN 0-07-021382-8
    1. Sound studios—Design and construction.   2. Architectural
acoustics.   I. Title.
    TH1725.E8624   1997
    621.389'3—dc20   96-27586
                                                                        CIP

McGraw-Hill

A Division of The McGraw·Hill Companies

1 2 3 4 5 6 7 8 9 0   DOC/DOC   9 0 1 0 9 8 7 6

ISBN 0-07-021382-8

The sponsoring editor for this book was Steve Chapman, the editing supervisor was Fred Bernardi, and the production supervisor was Pamela A. Pelton. It was set in ITC Century Light by North Market Street Graphics.

Printed and bound by R. R. Donnelley & Sons Company.

McGraw-Hill books are available at special quantity discounts to use as premiums and sales promotions, or for use in corporate training programs. For more information, please write to the Director of Special Sales, McGraw-Hill, 11 West 19th Street, New York, NY 10011. Or contact your local bookstore.

Sound Studio Construction on a Budget

To Elva
Her children arise up, and call her blessed;
her husband also, and he praiseth her.
Ecclesiastes

Contents

Introduction

DIFFUSION HAS DESCENDED ON THE SOUND RECORDING AND reproduction world. The last ten years could be called the diffusion revolution; Schroeder published his brilliant acoustical ideas from number theory in 1975, and D'Antonio has been busy since 1984 putting the acoustical theory into hardware for our studios. Reflection and absorption of sound must now move over to accommodate diffusion in the design and construction of sound rooms of all kinds.

Another new thing is the appreciation of the effect of early reflected sound on perceived sound quality. A good listening "sweet spot" today is one in which early sound, bounced from walls, ceiling, and floor, is controlled. Comb filter distortion, resulting from the interaction of early reflections with the direct sound, might be a distortion that is not all bad. It gives the music a sense of spaciousness, enveloping the listener in the music. It makes a small room seem larger; it gives breadth and stability to the stereo image.

These are all good things for most listeners, but the professional listener at the control room console wants to get rid of all distortion so that the signal can be evaluated with microscopic precision. The ten studio designs in chapters 1 to 10 take full advantage, where applicable, of diffusion of sound and the control of early reflections. In this sense, these ten designs point the way to a new era in sound recording and enjoyment of reproduced sound. This book follows essentially a nonmathematical approach to the design of spaces for both commercial and consumer use in recording and reproduction of sound.

The emphasis is on the practical, but theory is not far behind in guiding and explaining the practical. The last eight chapters are theoretical support for the first ten construction chapters. The practical reader's "good stuff" is put up front, and the professor's important "good stuff" follows.

Projected readership of this book includes students in courses related to music recording and reproduction, and music technology; musicians interested in sound recording; production and mixing personnel of recording studios; hi-fi enthusiasts interested in improving their listening room, or in building their own home theater; and engineering types of all stripes who are just plain interested in sound.

F. Alton Everest
Santa Barbara

Sound Studio Construction on a Budget

Voice-over recording studio

"VOICE-OVER" RECORDING IS THAT PROCESS BY WHICH A narration is recorded to later be used alone, or mixed with background music and/or sound effects. It can also be applied to the simple recording of voice alone. The studio for doing this is called a "voice" studio, with its particular acoustical characteristics, even as a "music" studio has its own acoustical requirements. At first glance it might seem that the acoustical requirements for a voice studio would be simpler than those of a music studio. This is not necessarily true. Voice sounds are subject to colorations (such as those caused by modal resonances and comb-filter effects), which can also affect music, but which are less audible in music.

What size should the voice studio be?

Although economic factors can outweigh the acoustical, there is a penalty in having a too-small studio. The modal resonances of the room *are* the acoustics of the room (see chapter 14). If the room is too small, modal resonance frequencies will be too few, with too-great spacing between them. This becomes a permanent flaw of the room, with no satisfactory correction. The smaller the room, the higher the low-frequency limit; that is, the lowest frequency with resonance support. For example, a room ten feet long will have a bass limit of 56 Hz, but a 20-foot room will have a bass limit of 28 Hz. Above 300 Hz the modal frequencies are so close together that problems associated with the lower frequencies tend to disappear.

The acoustical engineers of the British Broadcasting Corporation, with their hundreds of studios, have (on the basis of voice coloration studies) decided that it is impractical to build voice studios smaller than 1,500 ft^3.

What shape should the voice studio be?

In this rectilinear world a rectangular shape is assumed, but what dimensional proportions should be selected? Scores of papers have been written presenting arguments about why certain room proportions give the most uniform distribution of room modes. All have strong and weak points; none result in the perfect distribution of modal frequencies. Here are three proportions that have stood the test of time:

	Height	Width	Length
A.	1.00	1.14	1.39
B.	1.00	1.28	1.54
C.	1.00	1.60	2.33

Assuming a ceiling height of ten feet, these three proportions offer the following:

	Height	Width	Length	Volume
A.	10.0 ft	11.4 ft	13.9 ft	1,585 ft^3
B.	10.0	12.8	15.4	1,971
C.	10.0	16.0	23.3	3,728

Proportion A gives a volume that has been classed as marginal in size for a room to be used for a voice studio. Proportion B is a bit better, but proportion C is selected as the most promising.

Axial-mode study of selected room

In chapter 11, the axial, tangential, and oblique modes are discussed, and the axial modes are identified as the most potent. An appraisal of the axial modes alone can give a quick judgment of the acoustical quality of the room. Table 1-1 lists the axial modes of the $10 \times 16 \times 23.3$ ft room.

The lowest axial mode can be found from $f = 1130/2L$, in which 1130 is the speed of sound in ft/sec and L is the length of the room, 23.3 ft: $f = 1130/[(2)(23.3)] = 24.2$ Hz. Integral multiples of 24.2 Hz extend up through the spectrum, but we stop at 300 Hz because modal problems are rare above that frequency. These are the characteristic frequencies associated with the standing waves between the front and the rear walls. Similar computations are made for the 16′ width and the 10′ height in Table 1-1.

■ Table 1-1 Axial-mode study (room: 23.3′ × 16.0′ × 10′)

	Length L = 23.3′ $f_1 = 565/L$	Width W = 16.0′ $f_1 = 565/W$	Height H = 10′ $f_1 = 565/H$	Arranged in ascending order	Diff.
f_1	24.2 Hz	35.3 Hz	56.5 Hz	24.2 Hz	11.5 Hz
f_2	48.5	70.6	113.0	35.3	13.2
f_3	72.7	105.9	169.5	48.5	8.0
f_4	97.0	141.3	226.0	56.5	14.1
f_5	121.2	176.6	282.5	70.6	2.1
f_6	145.5	211.9	339.0	72.7	24.3
f_7	169.7	247.2		97.0	8.9
f_8	194.0	282.5		105.9	7.1
f_9	218.2	317.8		113.0	8.2
f_{10}	242.5			121.2	20.1
f_{11}	266.7			141.3	4.2
f_{12}	291.0			145.5	24.0
f_{13}	315.2			169.5	0.2
				169.7	6.9
				176.6	17.4
				194.0	17.9
				211.9	6.3
				218.2	7.8
				226.0	16.5
				242.5	4.7
				247.2	19.5
				266.7	15.8
				282.5	0.0
				282.5	8.5
				291.0	24.2
				315.2	

Mean diff. = 11.64
Std. dev. = 7.1

The important thing about axial modes is their spacing. For this reason all three sets of axial modes are arranged in ascending order to study this spacing. The right column of Table 1-1 isolates mode spacings for evaluation. The mean spacing is 11.6 Hz, which looks good when compared to 25 Hz, above which modal distortions will likely be audible.

There is a coincidence at 282.5 Hz which probably will not be audible because it is so close to the arbitrary 300-Hz limit. There is another near coincidence at 169.5 Hz, which can possibly pose a greater potential problem. Otherwise the prospects are excellent, with a warning to be alert to possible coloration of voice signals around 170 Hz.

What acoustical treatment?

An elevation sketch of the proposed $10 \times 16 \times 23.3$ ft room is shown in Fig. 1-1. The sound diffusion suggested at the right end of the studio represents a firm decision to stress diffusion in this and following studio designs. Diffusing elements that are economical and effective are now available and will be included in this voice studio.

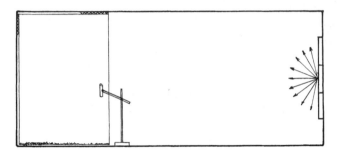

■ **1-1** *Elevation, voice studio.*

The left end of Fig. 1-1 is less definite. There are two approaches to treating the end of the room near the microphone to control the early reflections which are a major problem. Briefly, these reflections can be treated on an individual basis, or (taken together) all sound rays from the person at the microphone can be absorbed on a wholesale basis. There is a price to pay for this decision, which will be considered later.

The early reflections

The voice of the somewhat slinky individual at the microphone in Fig. 1-2 is reflected from nearby surfaces. Each first-order reflection impinges on a specific area of floor, side or end walls, and ceiling. Corner reflectors formed by the intersection of two or three surfaces return their energy directly back to the source.

The first reflection to return to the microphone is reflection 1 from the floor. It can be of significant magnitude because of the short path. A ceiling reflection, 2, is directly over the source. The three vertical walls around the source send reflections 3, 4, and 5

4

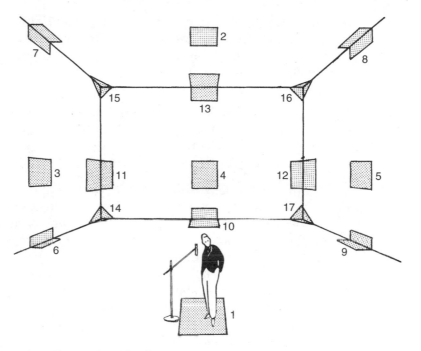

■ **1-2** *First-order reflections.*

back to the microphone. These five have undergone a single, normal (right-angle) reflection.

Corner reflections 6, 7, 8, 9, 10, 11, 12, and 13 are from the intersection of two planes, while 14, 15, 16, and 17 are from triplanes. Those from two-plane intersections undergo two reflections; those from triplane intersections undergo three reflections. As there is loss at each reflection, the corner reflections are of lower relative level than the normal reflections, and the triplane reflections are lower than the two-plane reflections.

The dominant early first-order reflections are numbers 1 through 5. They are of the greatest magnitude and earliest in time. The corner reflections are later and lower, but cannot be neglected in some circumstances.

The soft option and the hard option

The plan is to pick up the direct voice signal with the microphone of Fig. 1-1. After the direct voice signal arrives, reflections 1 through 5 arrive with good amplitude, and then 6 through 17 with reduced amplitude. After that, multiple bounces arrive from nearby walls and, later, from distant walls (including the far end of the room). Research has revealed clearly the effect of these early reflections on the direct signal (see Fig. 18-7). What happens during the first 60 milliseconds has a very great effect on the quality of the direct signal.

The "soft" option plan is to cover all surfaces of the entire left end of Fig. 1-1 with sound-absorbing material. This would surely reduce the amplitude of all early reflections. It can also increase the overall absorbence of the room too much, and make the room too dead.

The "hard" option plan would be to leave the entire left end of Fig. 1-1 with untreated surfaces except for the specific spots where reflections strike the walls, floor, or ceiling. A rug is placed under the microphone to reduce reflection 1. Reflections 2 through 5 are reduced by placing a $2' \times 2'$ absorber at spots where they hit room surfaces. This will be sufficient to clean up the direct signal. It probably will not be necessary to place absorbers at 6 through 17, but their presence is acknowledged. Reducing the amplitude of reflections 1 through 5 has been very successful in listening rooms. Referring again to Fig. 18-7, there is the question of whether the early reflections should be reduced below threshold or to line B, which would give a sense of spaciousness to the signal.

The soft option would tend to reduce all early signals below threshold (inaudibility). This is a tremendous improvement over having all reflections mixed with the direct signal, but is it the best? More work needs to be done to determine more precisely the extent to which early sound should be reduced.

Diffusion

Diffusion in Fig. 1-1 is obtained by the Skyline diffusor manufactured by RPG Diffusor Systems (Ref. RPG). Each $24'' \times 24''$ unit is made up of a 12×13 matrix of small blocks of thermoformed polymer to form a two-dimensional diffusor based on a primitive-root, number-theory sequence (see chapter 12). The right end of the studio of Fig. 1-1 supports a three-high by six-wide array of these $2' \times 2'$ omnidirectional diffusors.

These diffusors will form a solid basis for recording high quality voice signals from this studio once the design of the microphone end is decided. To settle that requires a bit more consideration of general studio characteristics.

General mode control

The best approach to providing general control of the low modal frequencies is to apply absorbing material or structures in room corners. For this reason, four RPG B.A.S.S. Traps are mounted on the far wall adjacent to the Skyline diffusors, two on each side. The absorption characteristics of the B.A.S.S. Trap are shown in Fig. 5-11. This membrane-type absorber offers attenuation in the low-frequency region required for normal mode control.

Voice studio design—soft option

Figure 1-3 is an "exploded" plan, so called because the four walls are laid down as though an explosion took place inside the studio. The basic construction of the space is assumed to be gypsum board on a wood frame, with a wood floor. It is assumed that the walls have adequate STC rating (chapter 13) to provide an ambient noise level within the studio of about NCB-15 (see Fig. 11-8). This will require close attention to the HVAC (Heating-Ventilating-Air Conditioning) equipment mounting and duct planning (see chapter 17). There is more to building a studio than meets the eye.

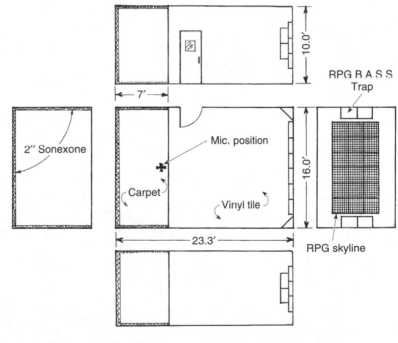

■ **1-3** *Soft option studio.*

Figure 1-3 is based upon an absorptive end around the microphone. The floor of this end is covered with heavy carpet and pad out to the 7' point. There are many types of sound absorption material suitable for the soft-end walls and ceiling. One possibility is 2" SONEXone, manufactured by Illbruck Acoustic Products (see Figs. 16-15, 16-16, 16-17). It must be remembered that the reflections are of mid to high frequency, and most sound-absorbing materials are effective in this higher range.

Voice studio design—hard option

Figure 1-4 specifies conditions for the hard option. The four B.A.S.S. Traps are mounted in the corners far from the microphone. The 18 Skyline diffusors are also mounted on the same end. An area rug is placed under the microphone and narrator, which is labeled 1 to conform to the absorber numbers in Fig. 1-2. The ceiling absorber, 2, is mounted directly over the narrator's position. Then come the side wall and end wall absorbers 3, 4, and 5. All of these absorbers are approximately 2' × 2', and 3, 4, and 5 are centered at the lip/microphone level, approximately 4' above the

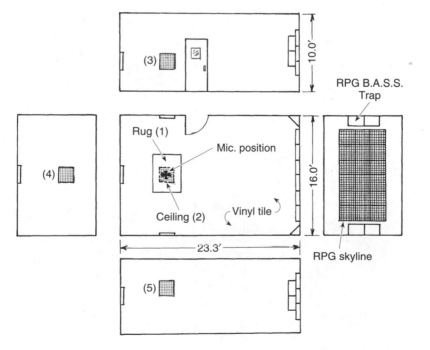

■ **1-4** *Hard-option studio.*

floor. This completes the specific units for the hard-option voice studio. Note that there is very little general absorption in the room. Will it be too live?

Reverberation time

A measure of liveness is reverberation time. Early in the development of acoustic arts and sciences, reverberation time was considered to be a most important measure of acoustic quality of a music hall. Today it is only one of many such indicators, and of much less importance. In such large spaces, ergodic (thoroughly mixed) conditions are approached and the Sabine reverberation equation applies. This equation can be stated as:

$$RT = \frac{0.049\, V}{Sa} \qquad (1\text{-}1)$$

in which:

RT = reverberation time, seconds
V = volume of space, ft^3
S = total surface area of room surfaces
a = Sabine average absorption coefficient
Sa = total absorption of space, sabins

A single absorption coefficient a that applies to all surfaces in a room is unlikely. Practically, many different materials with different coefficients are on the surfaces. A summation of the individual Sa products give the total sabin absorption in the room, from which the reverberation time is calculated according to equation 1-1.

The application of the Sabine equation to studio-sized spaces is somewhat controversial. Certainly ergodic conditions do not exist in the studios of Figs. 1-3 and 1-4. However, we need to be able to get started in blocking out a rough approach to treating a studio space. In this book, a limited application of equation 1-1 in such small spaces is approved, with full acknowledgment of its limitation. In other words, a reverberation time is computed as a tentative guide. The audible band is so wide (10 octaves) that the wavelength of sound in the upper registers is short enough that ergodic conditions are more closely approached in a studio. In the low frequencies, the decay of sound in the room involves chiefly the decay of a few modes. We can, therefore, depend more on the calculated high-frequency values of reverberation time of a studio than on the low-frequency values.

Reverberation time calculations

First, the reverberation time of the soft option of Fig. 1-3 will be calculated. The complete calculations are in Table 1-2, and it is advised that these be studied carefully to get an appreciation of where the sound is absorbed.

Section A of Table 1-2 covers the natural absorption of the room, which can very easily be overlooked. The gypsum board walls and the wood floor act as diaphragms, and in vibrating they absorb sound. The coefficients for gypsum board walls and wood floor are found in tables in the appendix. Separate calculations are made for each of the standard six frequencies for which coefficients are available. It is noted that the absorption in the room due to the bare walls and floor varies from 159.9 sabins at 125 Hz to 57.3 sabins at 4 kHz.

Section B of Table 1-2 applies to the surface treatments at the microphone end which give it the soft or absorbent characteristics. The absorption of SONEXone and carpet varies from 70.8 sabins at 125 Hz to 441.6 sabins at 4 kHz.

Section C of Table 1-2 applies only to the RPG B.A.S.S. Traps. Absorption coefficients have been gleaned from Fig. 5-11, but the area is so small the total absorption is small except at low modal frequencies. The absorption of the spot absorbers 1 through 5, mounted around the microphone to reduce amplitude of first-order reflections, is shown in Section D of Table 1-2.

Reverberation time—soft option

The room absorption that applies to the soft option in Table 1-2 is listed as the sum of sections A, B, and C (or the walls and floor "natural" absorption, the SONEXone, the carpet and the B.A.S.S. Traps). The total of sabins for these three sections is given in Table 1-2 for all six frequencies. The reverberation time calculated from equation 1-1 is plotted as the lower curve in Fig. 1-5. The curve flattens off at a reverberation time of about 0.3 seconds, which should be about right for this space. There seems little chance that anyone would want to go lower than 0.3 second. There is no simple way of increasing the reverberation time without decreasing the absorbing material around the microphone, and that is exactly what has been done in the hard option that follows.

■ Table 1-2 Reverberation time calculations (voice-over recording studio)

Material	S = area ft²	125 Hz a	125 Hz Sa	250 Hz a	250 Hz Sa	500 Hz a	500 Hz Sa	1 kHz a	1 kHz Sa	2 kHz a	2 kHz Sa	4 kHz a	4 kHz Sa
A. Walls + Ceiling	1039	0.10	103.9	0.08	83.1	0.05	52.0	0.03	31.2	0.03	31.2	0.03	31.2
Floor, wood	373	0.15	56.0	0.11	41.0	0.10	37.3	0.07	26.1	0.06	22.4	0.07	26.1
Sa, untreated			159.9		124.1		89.0		57.3		53.6		57.3
RT, untreated			(1.15)		(1.47)		(2.05)		(3.19)		(3.41)		(3.19)
B. 2″ SONEXone	412	0.15	61.8	0.34	140.0	0.81	333.7	1.00	412.0	0.92	379.0	0.90	371.0
Carpet, heavy	112	0.08	9.0	0.27	30.2	0.39	43.7	0.34	38.1	0.48	53.8	0.63	70.6
Sonex + carpet			70.8		170.2		377.4		450.1		432.8		441.6
C. B.A.S.S. Trap	16	0.4	6.4	0.2	3.2	0.15	2.4	0.1	1.6	0.05	0.8	0.05	0.8
D. Spot absorbers	20	0.02	0.4	0.22	4.4	0.69	13.8	0.90	18.0	0.96	19.2	1.0	20.0
Soft Design A + B + C			237.1		297.5		468.8		509.0		487.2		499.7
RT, sec.			(0.77)		(0.61)		(0.39)		(0.36)		(0.37)		(0.37)
Hard-End Design A + C + D			166.7		131.7		105.2		76.9		73.6		78.1
RT, sec.			(1.10)		(1.39)		(1.74)		(2.38)		(2.48)		(2.34)

Vol. = 3,728 cu ft

$$RT = \frac{0.049\,V}{Sa} = \frac{(0.049)(3728)}{Sa} = \frac{182.67}{Sa}$$

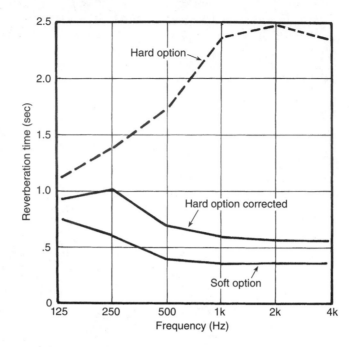

■ **1-5** *Reverberation time graph.*

Reverberation time—hard option

The room absorption in Table 1-2, which applies to the hard option, gives the total absorption as A + C + D with only the five 2′ × 2′ panels of 2″ absorbing material on the reflection spots (actually, the rug spot, 1, should be a bit larger). The sabin totals for each frequency are used to calculate the reverberation time according to equation 1-1. These reverberation times are plotted on the graph of Fig. 1-5 as the upper broken curve.

Reverberation time for this hard (more reflective) option range is from 1.10 sec. at 125 Hz to 2.34 sec. at 4 kHz. Few recording engineers will be happy with this too-bright condition, and it is assumed that it will be corrected. To achieve a reasonable reverberation time of about 0.5 seconds, additional absorption is required.

The rising reverberation curve of Fig. 1-5 for the hard option can be corrected after the need is verified by reverberation measurements and listening tests on voice signals recorded in the room. It will undoubtedly be judged as too reverberant. One possible plan for correcting it is mounting 86 Tectum (See Tectum Ref) sound blocks, 43 to each side wall. Each block is 15-1/2″ × 15-1/2″ by 2″

thick and is fabric covered. They should be mounted 24″ on centers to achieve the absorption figured into the "hard option corrected" curve of Fig. 1-5, and can be oriented as squares or diamonds.

Reverberation patterns

The reverberation pattern of Fig. 1-6 shows the time-level details of the hard option studio *before the spot absorbers are installed.* The reflections from spots 1 through 5 are labeled at their respective levels and delays. With the spot absorbers installed, these would be reduced 15–20 dB with respect to the direct signal. With the soft option treatment these reflections would be reduced about the same amount, 15–20 dB. In both options, there is a strong direct signal very close to zero delay. Also in both options, a nice 30 ms arrival time gap prevails before the late reflections and diffusion fall away.

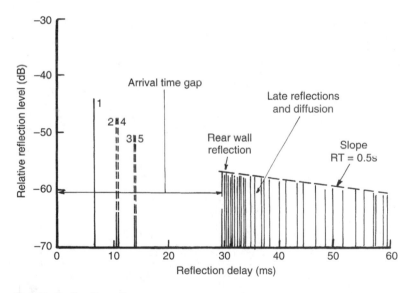

■ **1-6** *Reflection diagram.*

Concluding comments

The basic approach of this chapter has been to present an economical voice-over studio design that would give quality results. This has been done, and versatility has been included. Both the soft design of Fig. 1-3 and the hard design of Fig. 1-4 have excellent prospects. The five spot absorbers in the hard design reduce the amplitude of the first-order reflections, but do not affect the corner reflections. Do we want to wipe out these reflections, or do we want to adjust their amplitudes to some desirable spot on the research curve of Fig. 18-7? There is room for experimentation in the hard design, but all early reflections in the soft design are absorbed.

Recording studio for modern and classical music

2

A RECORDING STUDIO FOR MODERN AND CLASSICAL MUSIC must be large enough to accommodate many musicians. A music recording studio large enough to seat a full symphony orchestra and provide the acoustical space to do justice to such music is beyond the scope of this book. However, a music studio might be possible that is large enough to provide a proper acoustical environment for the recording of choral groups, chamber music groups, community symphony orchestras, and even modest-sized bands (although bands, in general, need brighter acoustics than the more classical music groups).

What size studio?

The goal is to design a recording space large enough to avoid the cramped conditions medium to large music groups encounter when forced to record in the average recording studio. Only the musicians are to be considered in this space. No audience area is to be provided, although officials and casual hangers-on can be seated informally.

The floor area and the volume of the space should be kept low enough to shorten the distance reflected rays must travel on their way to the microphone.

Reverberation time

Reverberation time is that time required for sound in a space to decay 60 dB. If surfaces are highly reflective, this time will be long. If the surfaces are very absorptive, reverberation time will be short, and sound will die away quickly. Reverberation time of music halls and auditoria has changed from being considered the most important criterion of quality to being considered simply one of the numerous criteria. The importance of reverberation time has been diminished, but it is still one of the factors to be studied in the design of music (and other) spaces.

The reverberation time that best serves the various types of music and speech is shown in the graph of Fig. 2-1. These graphs represent the opinions of many music experts, and must be approached with some skepticism, because the experts are not in full agreement. The "Concert Studio" graph is close to the type of space of interest for a recording studio for modern and classical music. Organ music thrives on longer reverberation time, while chamber music requires a more intimate acoustic associated with a shorter reverberation time. Speech, of course, fares best in spaces of short reverberation so that syllables will not be slurred together.

For a recording studio for modern and classical music of 100,000 ft^3 volume, a reverberation time of about 1.6 seconds is indicated. This will be taken as the target value for preliminary calculations.

16 Diffusion in the recording studio

Diffusion of sound (see chapter 12) has been considered to be of utmost importance in music halls from the earliest days. It has traditionally been achieved by geometric irregularities in the side walls and ceilings of the music spaces. All that was changed by the availability of Schroeder's reflection phase-grating diffusor, which is far more efficient than geometrical protrusions. In this recording studio for modern and classical music, diffusing elements are on three walls and the ceiling.

Studio design

The design offered for recording of modern and classical music is shown in Fig. 2-2, with a description of all materials in the key to the figure. The dimensions of the room are 70.0' × 48.0' × 30.0' which give a volume of 100,800 ft^3. The six surfaces of the room are shown in Fig. 2-2 with all major acoustical elements.

This is considered as a stand-alone building, with a control room housing to be added on the south end of the building. Control rooms will be considered in chapter 6.

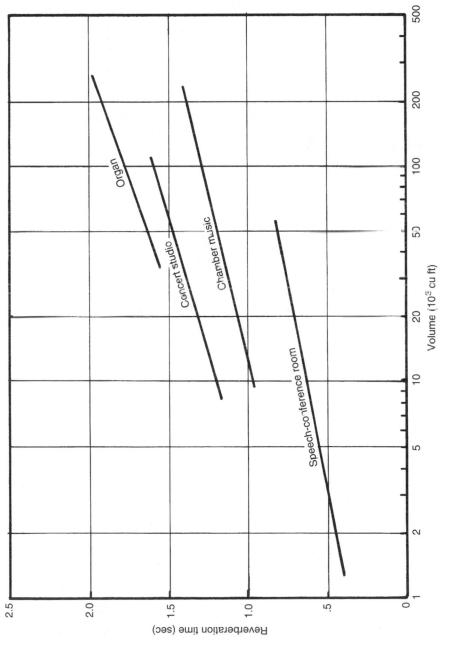

■ **2-1** Optimum reverberation time.

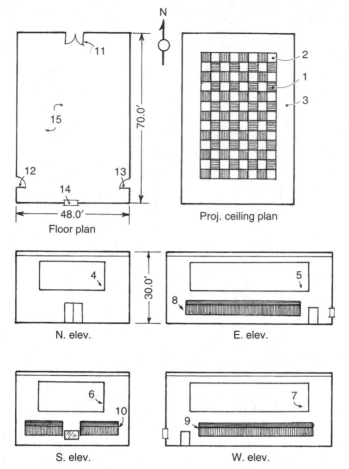

(1) RPG Abffusor, 4′ × 4′, quantity 52. (2) RPG NDC Almute, 4′ × 4′, with 2″ × 4′ × 4′ Owens-Corning 703. (3) Tectum Designer, 1-1/2″ × 24″ × 48″, E-400 Mtg., quantity 100. (4–7) Tectum Fabri-tough wall panel, C-40 Mtg.: (4) 12′ × 28′, (5) 12′ × 50′, (6) 12′ × 28′, (7) 12′ × 50′. (8–9) RPG QRD 734, each 2′ × 4′, quantity 36, rack mtd. (see Fig. 2-3). (10) RPG QRD 734, each 2′ × 4′, quantity 24, rack mtd. (see Fig. 2-3). (11) Overly swinging door pair, 2-1/2″ thick, STC-50. (12–13) Overly single swinging door, 1-3/4″ thick, STC-50. (14) Control room observation window, 4′ × 6′ (see chapter 15). (15) Floor, wood parquet on concrete.

Floor plan

Proj. ceiling plan

N. elev.

E. elev.

S. elev.

W. elev.

■ **2-2** *Plan and elevation views, music studio.*

The building is basically a concrete-block structure with double-leaf walls for protection from outside noise, and pillars to support the roof structure. The pillars are flush on the inner walls. The actual specification of the walls is to await an evaluation of the exterior noise exposure, but an STC-50 (Sound Transmission Class) classification for the walls is considered a minimum (see chapter 13).

A large door on the north end is wide enough for a light truck to enter for transport of instruments and equipment. It is closed by a double swinging door of STC-50 rating to match the walls. Single swinging doors on the south end of both the east and west walls are rated at STC-50.

Acoustical treatment

The unique acoustical features are the diffusors on the wall and in the suspended ceiling. The wall diffusors are built around the QRD-734 supplied by RPG Diffusor Systems, Inc. These diffusors are described in chapter 12, and an exploded view of a single unit is shown in Fig. 2-3. This basic quadratic-residue diffusor of prime 7 is fitted with top and bottom plates, and with end plates as required.

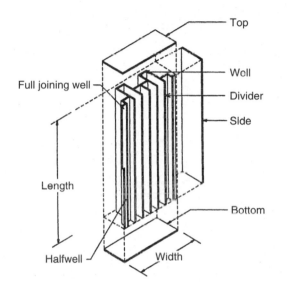

■ **2-3** *QRD-734 construction.*

A basic cluster of three such 2′ × 4′ units is shown in Fig. 2-4. The top horizontal unit diffuses sound in the vertical direction, and the lower units with vertical wells diffuse sound in the horizontal di-

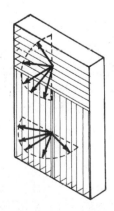

■ **2-4** *Diffusion of three-group 734s.*

rection. Between the two, full two-dimensional diffusion is accomplished.

The racks of 734s on the walls are supported on 3/4″ × 3″ wall cleats as shown in Fig. 2-5. A 48′ row of 734s is placed on the east wall and on the west wall. A 16′ row is placed on either side of the control room observation window on the south wall. A total of 96 individual 2′ × 4′ QRD 734s are used.

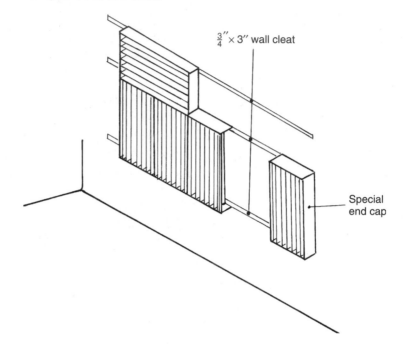

$\frac{3}{4}″ \times 3″$ wall cleat

Special end cap

■ **2-5** *Wall mounting of 734s.*

The placement of the rows of 734s on the walls is critical to prevent flutter echoes between opposing, parallel, reflecting surfaces. Only horizontal rays of sound can excite flutter echoes. If all sound sources (such as musical instruments) are 2′ to 4′ above the floor, the bottom of the diffusors should be about that distance from the floor. In Fig. 2-2, diffusors 8, 9, and 10 are shown 4′ from the floor. It is possible that these should be lowered, but there are so many scatterers close to floor level between the two reflecting surfaces that flutter echoes that low are unlikely.

Wall panel absorbers

For general absorption, wall panels 4, 5, 6, and 7 of Fig. 2-2 have been added. These could be made of any one of a dozen brands of commercial absorbing panels, but Tectum Fabri-Tough panels of 1″ thickness have been selected. These panels are installed with the standard C-40 mounting, which means with an air space of 40 mm (1-1/2″) between the panels and the wall. This improves the low-frequency absorption materially. The absorption of these panels will be about 30% at 125 Hz, 77% at 250 Hz, and essentially 100% through 4 kHz. Should measurements and experience show that the reverberation time of the studio should be lowered or increased, removing or adding wall panel area is a practical approach.

Ceiling acoustical treatment

The suspended ceiling is the standard C-400 mount, in which the ceiling is dropped down 400 mm (16″). This ceiling is of the type illustrated in Fig. 2-6; it serves multiple functions. In addition to HVAC (heating-ventilating-air-conditioning) ducting, fluorescent lighting units could be accommodated. However, in this studio drop lighting units are considered more desirable as they bring the light source closer to music stands, etc.

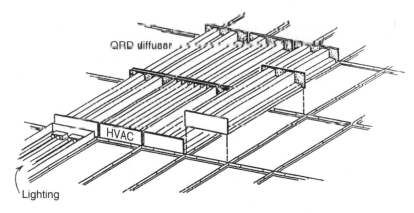

QRD diffusor

HVAC

Lighting

■ **2-6** *T-bar mounting of units.*

Both diffusing/absorbing and ordinary absorbing units are mounted in the suspended ceiling frame. The RPG Abffusor is both a diffusor and an absorber. The Abffusor combines the high/midfrequency properties of porous materials with the low-frequency properties of diaphragmatic membranes to give good absorption down to 100 Hz for all angles of incidence. The absorption coefficients of the Abffusor and other materials used are listed in Table 2-1.

There are three lay-in panels in the suspended T-frame: one is the Abffusor and the other is an NDC Almute panel with 2″ of 6 lb/ft^3 glass fiber (Owens Corning 705) on it. The Almute panel is a

2.5-mm (3/32″) panel fabricated from sintered aluminum particles with many 100-micron holes. The glass fiber backing should not touch the panel. The 52 Abffusors and 52 Almute panels fill the central portion of the suspended ceiling.

The third type of lay-in panel surrounds the central portion of the suspended ceiling. These are 424 panels of 2′ × 2′ × 1-1/2″ Tectum Designer panels. This 8- to 10-ft band of Tectum around the edge of the ceiling places the diffusing elements nearer the center of the room for greater effect.

Reverberation time

The reverberation time of this recording studio must be verified by measurements. Calculations can give only a rough idea of reverberation time because of the many factors that can enter between theory and practice. Table 2-1 is a rough assemblage of the various contributions to the absorption of the room. It is broken down into the suspended ceiling absorption and the wall absorption.

The absorption coefficients a of the different materials are multiplied by the surface area S of each material to obtain the Sa product for each material for the six frequencies. The Sa product is the number of absorption units (sabins) contributed by each material. The sum of the absorption units is added to find the total absorption units for each frequency. Sabine's equation for reverberation time is:

$$\text{Reverberation time} = \frac{0.049\,V}{Sa} \qquad (2\text{-}1)$$

in which:

V = volume of the space, ft^3
Sa = total absorption of the space, sabins

For this particular space, equation 2-1 becomes:

$$\text{Reverberation time} = \frac{(0.049)(100{,}800)}{Sa}$$

$$= \frac{4939}{Sa}$$

The concrete block walls are painted, resulting in minor (but not to be neglected) sound absorption changes.

■ Table 2-1 Reverberation time calculations

Material	S = area ft²	125 Hz a	125 Hz Sa	250 Hz a	250 Hz Sa	500 Hz a	500 Hz Sa	1 kHz a	1 kHz Sa	2 kHz a	2 kHz Sa	4 kHz a	4 kHz Sa
Suspended ceil.													
RPG Abffusor	208	0.82	171	0.90	187	1.00	208	1.00	208	1.00	208	1.00	208
NDC Almute + 2" 703	208	0.90	187	0.92	191	0.99	206	1.00	208	1.00	208	1.00	208
Tectum design.	1696	0.44	746	0.47	797	0.36	671	0.51	865	0.71	1209	1.00	1696
Wall treatment													
RPG 734s	768	0.23	177	0.24	184	0.35	269	0.23	177	0.20	154	0.20	154
Tectum Fab. Tch. C-40	1872	0.30	562	0.77	1441	1.00	1872	0.98	1835	0.79	1479	0.95	1778
Conc. blk. paint	4288	0.10	429	0.05	214	0.06	257	0.07	300	0.09	386	0.08	343
Floor, parquet on concrete	3360	0.02	67	0.03	101	0.03	101	0.03	101	0.03	101	0.02	67
Total Sa, sabins			2339		3115		3584		3694		3740		4454
Rev. time, sec = 4939/Sa			(2.11)		(1.59)		(1.38)		(1.34)		(1.32)		(1.11)
Air absorption = rel. hum. 50% 100,800 cu ft								0.9/k	91	2.3/k	232	7.2/k	726
Total Sa with air			—		—		—		3785		3972		5180
Rev. time with 50% rel. hum.			(2.11)		(1.59)		(1.98)		(1.30)		(1.24)		(0.95)

$$\text{Rev. time} = \frac{(0.049)(100{,}800)}{Sa} = \frac{4939}{Sa}$$

Calculations

Table 2-1 shows the calculations for the 100,800 ft³ music recording studio that has been described. It is first calculated without air absorption, which is added on later. The plotted results of these calculations are shown in Fig. 2-7. Single-figure appraisals of reverberation time are commonly made at 500 Hz.

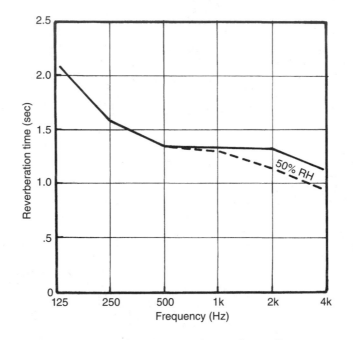

■ **2-7** *Reverberation time graph, music studio.*

The 500-Hz reverberation time in this case is about 1.4 seconds. Referring to Fig. 2-1 shows that a "concert studio" of 100,000 ft³ size calls for a reverberation time of about 1.6 seconds. Chamber music, on the other hand, calls for a reverberation time of about 1.3 seconds. With the uncertainties of different types of music as well as calculations vs. measurements, the preliminary value of 1.4 seconds is reassuring.

Air absorption

In a structure of this size, absorption in the air must be considered. Its effect is only at the high-frequency end of the audible spectrum, but it can seriously rob the signal of important high-frequency content. Modern air conditioning tends toward stabilizing this effect.

Overview

Reverberation time is a slender reed on which to lean. The present design actually depends on reverberation time only to arrive at a reasonable distribution of absorbing and diffusing elements. The dependence on diffusion in this design is great. Generally diffuse conditions should result from the ceiling and the walls, and this is expected to be the main characterization of this studio.

What is the effect of the diffusors? Musicians should enjoy excellent ensemble, that is, the individual musician would be able to hear the other musicians, which is necessary for coordination. Music in a diffuse space has an open, free, uncluttered character that the experts seek. The individual in a well-diffused small space, such as a control room, has the impression of being in a larger space.

The arrival time gap

In recorded music the arrival time gap is the short span of time between the arrival of the first component of the direct sound and the reflected components arriving shortly thereafter. This gap is imprinted on every music signal and has an important effect on the quality of the music. Experts in the fields of music and music-hall acoustics tend to agree that the arrival time gap is related to the subjective impression of the "intimacy" of the sound. Those reflections in the concert hall arriving laterally have been demonstrated to be related to the "spaciousness" of the music. Thus subjective impressions such as "clarity," "warmth," and "brilliance" are slowly being related to specific acoustic measurements.

In the 1960s, Beranek (Ref.) used the term "initial time delay gap," which is still in use today. In the present context the shorter and more descriptive term "arrival time gap" will be used.

In a concert hall a person seated in the audience first hears the onset of the direct sound, followed by reflections from the sidewalls, ceiling, and so on. The revered music halls of Europe and America have arrival time gaps in the range of 20–30 milliseconds.

The rough sketch of Fig. 2-8 is based on the dimensions of the present music recording studio. Taking a spot in the orchestra as the location of the source, the direct sound travels a 10-ft path to the microphone. Reflections from the ceiling travel about 41 feet

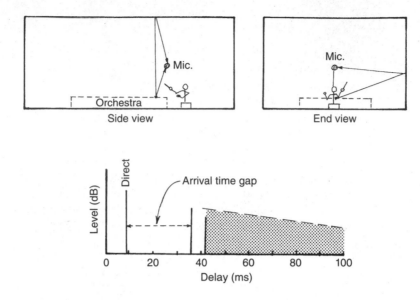

Side view   End view

■ **2-8** *Development of arrival time gap.*

to reach the microphone. Taking time = 0 as the time the sound leaves a spot in the orchestra, the direct 10-ft path takes 10 ft/ (1330 ft/sec) = 9 ms for the sound to reach the microphone. The reflection takes 41/1130 = 36 ms to reach the microphone. The arrival times of the direct and reflected components may then be plotted on the graph of Fig. 2-8 at 9 and 36 ms, respectively. The side wall reflections may be plotted beginning at 43 ms. The arrival of the direct ray opens the gap and the arrival of the first strong reflections closes the gap. The gap will not be "silent," but will be filled with a low-level clutter. As long as this clutter is 20–30 dB down from the peaks, the gap will serve its intended purpose.

Recording studio for rock music

3

THE SECRET OF A SUCCESSFUL ROCK MUSIC RECORDING studio is not in its acoustical treatment, it is in keeping the neighbors happy. The high sound levels surrounding the studio and control room must be kept within bounds, or the consequences must be suffered. "Pollution" is a power word these days, and among the general population are those who would jump at the chance of applying the word to rock music.

Precautions must be taken to avoid disturbing neighbors. Containing the high-level sounds of rock music is a very real technical challenge. If the sounds in the control room reach 120 decibels (near the threshold of pain), a 50-dB wall reduces the music to 70 dB which is about the level many people would operate their radio. This would not be so bad if the person *wanted* to hear rock music, but if he or she didn't? For this reason, protecting people around the rock recording studio is a first priority in this proposed studio.

Assumptions

We shall assume that the space to be used for the studio is on the ground floor of an existing concrete building. It is a corner space; the studio has two exterior walls and two interior walls, and there are neighbors on the same floor and on the floor above.

The control room is acknowledged only by placement of an observation window in the south wall. That is a separate project and will be carried no further in this chapter. Control rooms are considered more fully in chapter 6.

It is also assumed that a rock music studio would have need for a drum cage and vocal isolation booth. Lighting and HVAC are very important, but are not covered here. In other words, this is an acoustical design covering surface treatment, sound locks, and such.

Floor plan

The general floor plan is shown in Fig. 3-1. It includes a drum cage, an equipment storage room, a vocal booth, a baffle storage nook, and a sound lock that opens into the control room as well as the exterior. A control room window is included in the south wall, but the control room itself is another project. Approximately half the floor area is carpeted, with wood parquet on the other half.

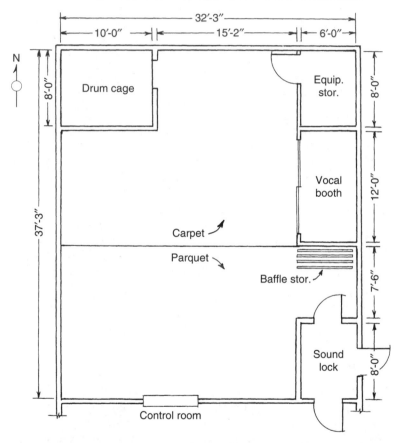

■ **3-1** *Plan view of a rock studio.*

Wall sections

Figure 3-2 locates sections D, E, F, and G (sections A, B, and C are reserved for drum cage walls, which are to be discussed later). Sections D and G are exterior walls, E and F interior. The inside walls for the drum cage, equipment room, and vocal booth are not indicated pictorially but are simple single layers of 5/8″ gypsum board on each side of metal channels.

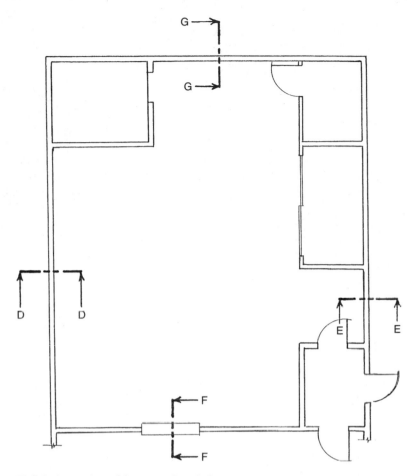

■ **3-2** *Location of four sectional views.*

Section D-D

Sections D and E are both shown in Fig. 3-3. The floor of 4″ concrete, common to both, is floated on compressed glass fiber or neoprene "hockey pucks." The concrete is not poured until the 3/4″ plywood form is in place and covered with a plastic sheet to keep concrete from seeping through cracks and forming a solid bridge to the supporting structure. Glass fiber perimeter boards are also to be shielded by plastic sheet. The concrete floor must be completely isolated from the structure to be effective. There must also be just enough of the pucks to bear the weight of the concrete itself, the walls, and the ceiling and be deflected about 15%, so that the full resiliency of the pucks can be realized.

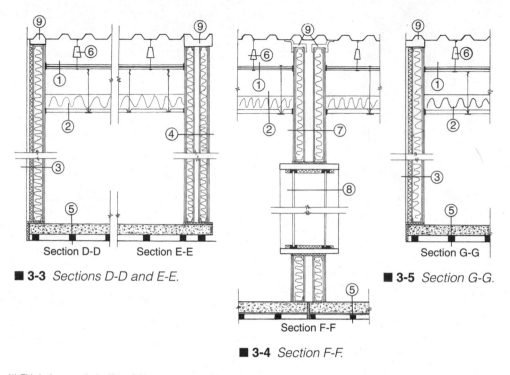

Section D-D   Section E-E

■ **3-3** *Sections D-D and E-E.*

Section G-G

■ **3-5** *Section G-G.*

Section F-F

■ **3-4** *Section F-F.*

(1) This is the acoustical ceiling of this space, composed of two layers of 5/8″ gypsum board separated by a dense vinyl septum. (2) Suspended T-frame ceiling, a vital element of acoustical treatment of the space, supporting Tectum Ceiling Tile (1″ × 24″ × 24″) with 6″ glass fiber backing. (3) Exterior wall, left to right: 2″ space, 1.5″ F.G. fill, 5/8″ gypsum board, 4″ building insulation, double 5/8″ gypsum board. (4) High-insulation inner wall, left to right: double 5/8″ gypsum board, 4″ building insulation, single 5/8″ gypsum board, 2″ air space, single 5/8″ gypsum board, 4″ building insulation, double 5/8″ gypsum board. (5) Floating floor; compressed F.G. or neoprene resilient blocks, 3/4″ plywood, plastic membrane, 4″ concrete floor. F.G. perimeter board to insulate floor from structure. (6) Resilient isolation hanger (see Figs. 14-5, 14-6). (7) Double internal wall, same as #4 except with 6″ spacing. (8) Double- or triple-glass observation window with heavy panes of different thicknesses. Glass panes isolated from frame. (9) Special wall stabilizer designed to isolate the wall from the structure. *Note:* All joints to be sealed with (and all metal wall runners to be set in) acoustical sealant.

All the walls of the recording studio are supported on the floating floor. The acoustical ceiling (1), hung from the structural building on resilient hangers (6), is made of two 5/8″ gypsum board layers separated by a septum of dense vinyl. This is to get a maximum mass for maximum protection of occupants on the second floor. The floating floor, the four walls, and this acoustical ceiling constitute a space within a space for maximum attenuation of the music within.

From the acoustical ceiling (1) hangs the usual suspended T-bar ceiling with a 16-inch air space to conform to the standard C-400 (400 mm) mounting. Layer 2 is composed of 1″ × 24″ × 24″ Tectum Ceiling Tiles laid in the T-frame with 6″ of glass fiber resting on the top of the tiles.

Wall 3 (Fig. 3-3) is built alongside the concrete west wall of the building with a 2″ spacing. This spacing is almost filled with 1.5″

glass fiber. The layer of the new wall is 5/8″ gypsum board attached to steel studs and channels. The inner space of the wall is filled with low-density building insulation, and the inner layer is a double layer of 5/8″ gypsum board. It will be mentioned now and repeated many times that all cracks and joints should be filled with acoustical sealant, the goal being to make the inner space hermetically independent. Cracks allow sound to escape, defeating the purpose of all the carefully designed walls, floor, and ceiling.

For common sense to be kept abreast of theory, it should be emphasized that resilient hangers (6), floating floor, and spacing wall (3) from the wall of the building are all to isolate the studio space from the structure. The reason is that the concrete structure is an excellent transmitter of noises both into and out of the studio. The major justification for going to all this expense is to protect the neighbors from the music.

Section E-E

The only feature of Section E-E (Fig. 3-3) not discussed in Section D-D is double wall (4), an inner wall that runs the length of the studio. It is simply a double wall spaced 2″. Each leaf of the double wall is composed of the standard metal studs and channels with a single 5/8″ gypsum board layer on the inside and a double layer of the same on the outside. Although not specified, it would be a good idea to squeeze out the last few dB of the STC rating by mounting the outer layer of 5/8″ gypsum board facing the other occupants on resilient channels.

Section F-F

Section F-F is shown in Fig. 3-4. A double- or triple-glass observation window having heavy (or laminated) panes of different thicknesses, each isolated from the frame by rubber edging. The frame, in turn, is isolated from the structure by pads of compressed glass fiber (see chapter 15).

Section G-G

Section G-G is shown in Fig. 3-5. All features of this section have been covered previously.

Drum cage

An isometric sketch of the proposed drum cage is shown in Fig. 3-6 and details of construction in Fig. 3-7. It is 8' × 10' × 11' with openings on south and east sides to give maximum view of the floor activity. Basically it features a wood floor of 1.5" tongue and groove decking on top of a 2" × 8" structure deadened by loading with sand. The entire floor is isolated from the building structure by hockey pucks such as used to float the floor. This gives the drummer the required solid floor which is essentially nonresonant.

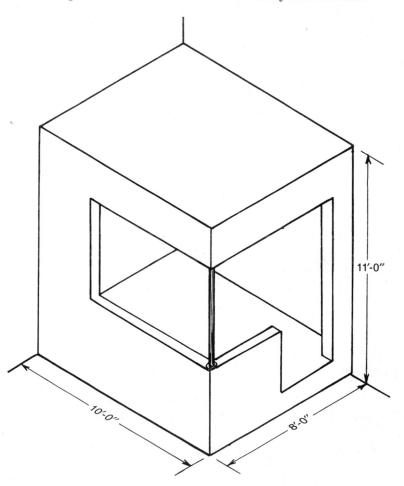

■ 3-6 *An isometric sketch of a drum cage.*

The whole purpose of the drum cage is to reduce the high sound levels from the kit at the position of the other instruments. To assist in this the ceiling of the cage is made highly absorbent by

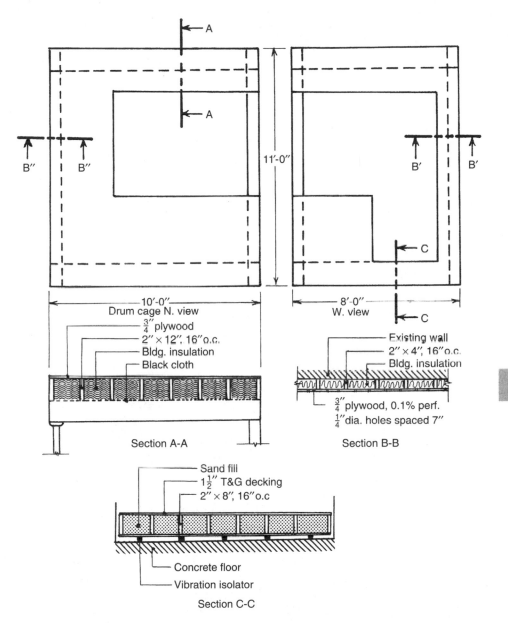

■ **3-7** *Drum cage details.*

packing the 2″ × 12″ frame with building insulation. The north and west walls of the cage are bright, reflecting surfaces of 3/4″ plywood, but made into a tuned structure by 1/2″ holes spaced 7″. These two walls have a tuned absorption peaking in the region of 80 Hz, but little absorption above 150 Hz. This should help keep the kick drum sound under control, yet provide the drummer with adequate personal return.

Vocal booth

The vocal booth is a $6' \times 12'$ space with a sliding glass door opening to the studio, which provides an excellent view both ways. This sliding glass door, when closed, will provide about 20 to 25 dB of insulation in addition to the inverse-square (or "6dB/distance-double") law. The acoustical treatment of the interior has not been specified, but something should be added to control axial modes. Absorption in the corners would be best for this.

Studio treatment—north wall

The north wall is covered with 2" Tectum panel with C-40 mounting, as indicated in Fig. 3-8. This means that the panels are furred out from the wall on 2×2 strips (net about 1-1/2") 24" on centers. This modest spacing from the wall improves the absorption of the panel so that it is essentially perfect above 250 Hz.

Studio treatment—south wall

A recording studio can be excellent acoustically, and very poor aesthetically. Some might associate drabness with poor sound quality. Musicians are aesthetes, and it is a fair guess that they, their guests, and clients would appreciate one wall of the studio devoted to an artistic piece. For this reason, the south wall of the studio (Fig. 3-8) would be an ideal place for an artistic panel of the type offered by Brejtfus (see Ref.). It could be a large scenic or a 20-foot dragon or a catchy geometric design. Brejtfus not only can make it, they can make it absorb sound. Their Artistic Sound Panel is a glass fiber base with a 1/8" high-density glass fiber layer on which a polyfoam layer bears the design.

Studio treatment—east wall

Broken up by many doors and other openings, the east wall can handle only a panel 4' to 6' in width running almost the length of the studio (Fig. 3-9).

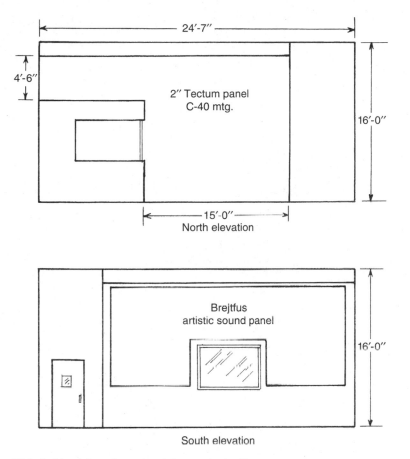

24'-7"

4'-6"

2" Tectum panel
C-40 mtg.

16'-0"

15'-0"

North elevation

Brejtfus
artistic sound panel

16'-0"

South elevation

■ **3-8** *North/south ends of the rock studio.*

Studio treatment—west wall

An array of six doors, 10' long and 24" wide, take up 24' of the west wall (see Fig. 3-9). Each door is hinged, and is absorbent on its inner side and reflective on the outer side. When each door is open it not only exposes its own absorbent backside, but also reveals another absorbent area the size of the door. When all doors are swung open, the entire 10' × 24' area is absorbent.

The ultimate purpose of the doors? It is not to alter the reverberation time, because this 240-square-foot area has only a minor effect on the overall reverberation time. It is basically to make available reflective and absorbent areas so that musicians can set up near one or the other to achieve the effects they desire. Note that the Tectum wall on the north end of the studio is absorbent,

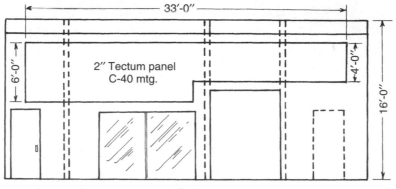

East elevation

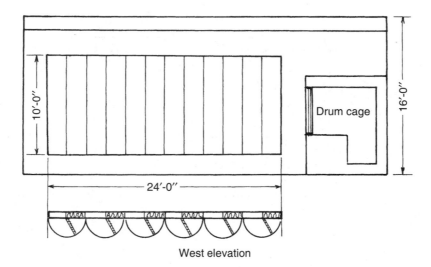

West elevation

■ **3-9** *East/west ends of the rock studio.*

but this area of swinging panels is the only other absorbent area close to musicians.

Studio treatment—floor, ceiling

The floor is about half carpet and half parquet wood, as shown in Fig. 3-1. A beautiful curved line of demarcation between the two would look better, but this line should be drawn by the occupants-to-be.

The suspended ceiling is a very absorbent surface, even if the space behind is shared with air-conditioning ducts and electrical equipment.

Sound lock corridor

The sound lock corridor provided in Fig. 3-1 might not be sufficient if neighbors are close to the outer door. Anyone in a hurry exiting from either the control room or the studio would go quickly through two doors, one leading to a high-sound area and the outer door to the protected area. A burst of sound could exit with the person. It might be advisable to devote more space to this sound lock.

All inner surfaces of the sound lock corridor should be very absorbent. The ceiling could be a suspended one using the same Tectum Ceiling Tile. The floor should be heavily carpeted. The walls could be the same furred out (C-40) Tectum wall panel. Tectum is made of wood fibers and is quite resistant to abrasion. The doors of the sound lock should be Overly Model STC5089108, which offer an STC 50 rating.

Reverberation time

There is little interest in the reverberation time of a rock recording studio. It matters little how fast or slow the room sound dies away, because multichannel recording demands close microphones or direct feeds to achieve sufficient channel separation. Reverberation time calculations are of interest to balance the absorption of different areas and materials. Actually, the goal is to make the studio as dead acoustically as possible. In the present rock recording studio design almost every available ceiling and wall area has been covered with highly absorbent material. In other words, it has been made as dead as possible for a room of 15,000 ft^3 volume. Our interest in reverberation calculations is then primarily to inspect the distribution of sabins of absorption about the room. Table 3-1 shows all the reverberation time calculations for each absorbent unit contributing to the overall acoustic of the space.

For each of the six standard frequencies the absorption coefficient a for a given material is multiplied by the area S of that material to obtain the Sa product in sabins. Adding all these unit absorbances in sabins for a given frequency gives the total absorption for that frequency from which the reverberation time may be estimated from the equation:

$$\text{Reverberation time} = \frac{(0.049)(\text{volume})}{Sa} \tag{3-1}$$

in which

■ Table 3-1 Reverberation time calculations

Material	S = area ft²	125 Hz		250 Hz		500 Hz		1 kHz		2 kHz		4 kHz	
		a	Sa	a	Sa	a	Sa	a	Sa	a	Sa	a	Sa
Suspended ceiling C-400 (16") TECTUM ceil. tile 1" × 24" × 24" lay-in 6" fiberglass back.	956	1.01	965	0.89	850	1.06	1013	0.97	927	0.93	897	1.13	1680
East wall panel: TECTUM wall panel C-40 mounting, 1½"	164	0.42	69	0.89	146	1.19	195	0.85	139	1.08	177	0.94	154
West wall adjust. panels:													
Fully opened	240		240		240		240		240		240		240
Fully closed	0		0		0		0		0		0		0
South wall panel, Brejitus Artistic Panel.	210	0.16	34	0.47	99	1.10	231	1.14	239	1.05	221	1.04	218
North wall fully covered with TECTUM wall panel C-40 mounting, 1½"	252	0.42	106	0.89	224	1.19	300	0.85	357	1.08	272	0.94	237
Drum cage ceiling	80	1.0	80	1.0	80	1.0	80	1.0	80	1.0	80	1.0	80
North & West walls 0.1% perf, 4" deep	128	0.8	102	0.3	38	0.2	26	0.15	19	0.15	19	0.1	13
Floor, heavy carpet	448	0.08	36	0.24	108	0.57	255	0.69	309	0.71	318	0.73	327
Floor, parquet	465	0.02	9	0.03	14	0.03	14	0.03	14	0.03	14	0.02	14
TOTAL			1641		1799		2354		2324		2238		2365
Reverb. time with West doors open, sec.			(0.45)		(0.41)		(0.32)		(0.32)		(0.33)		(0.32)
Reverb. time with West doors closed, sec.			(0.53)		(0.48)		(0.35)		(0.36)		(0.37)		(0.35)

$$\text{Volume of space, cu ft} = 15{,}220 \text{ cu ft}$$

$$\text{Reverberation time} = \frac{(0.049)(15{,}220)}{Sa}$$

$$= \frac{745.78}{Sa}$$

The calculated reverberation times have been plotted in Fig. 3-10 for the swinging panels both open and closed. These panels will probably never be swung for the purpose of adjusting reverberation time. The adjustment of these panels is primarily to adjust local acoustics to please musicians. It will take a keen, experienced ear to sense the difference between the 500-Hz reverberation times of 0.32 and 0.35 seconds brought about by swinging panels. A musician can easily sense playing alongside a hard versus a soft wall.

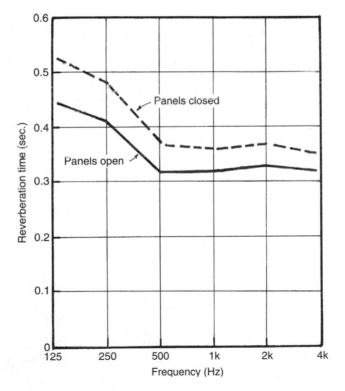

■ **3-10** *A reverberation time graph of a rock studio.*

A thought on baffles

The traditional baffle used to improve the separation between instruments has one reflective side and one absorptive side. The baffle suggested in Fig. 3-11 introduces diffusion of sound into the baffle technology. Sound returned from the 4′ × 4′ diffusor area is different from that returned from a flat reflective surface in the following ways: One, the return is about 8 dB lower in intensity; two, the return is greatly diffused through the half-space; and three, the return is spread over several milliseconds of time. The musician will undoubtedly sense a new roundness and full-

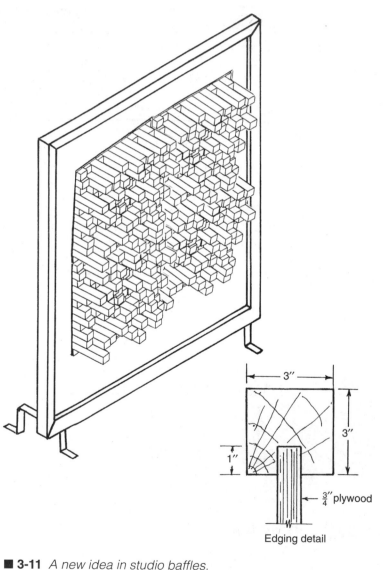

3″

3″

1″

$\frac{3}{4}$″ plywood

Edging detail

■ 3-11 *A new idea in studio baffles.*

ness to the sound. Whether it will be judged helpful remains to be seen.

The baffle in Fig. 3-11 has mounted on it four Skyline diffusors manufactured by RPG Diffusor Systems. It is suggested that Fig. 12-11 be carefully studied to understand the functioning of these diffusors.

4

Personal project studio

ONE OF THE FIRST THINGS AN ASPIRING MUSICIAN WANTS IS a studio in which demonstration recordings can be made to tell the world a new star has arrived. With limited resources, the logical location for such a studio is the living room of his home, or "that room out back." From that point on the musician is completely on his or her own, because the focus of technical books and periodicals that might be of help is usually on the larger, more expensive, professional studio.

This musician with recording and mixing ambitions might already have on hand a sound recorder envisioned as the heart of the new studio. In addition, a small mixing console, midi, drum machine, etc., might be on hand or available. From here on, only the big mystery of that intimidating word, *acoustics*, remains to be confronted.

Home acoustics—reverberation

The average reverberation time of fifty living rooms has been reported by Jackson and Leventhall (Jackson 1972) as being about 0.7 second at 100 Hz, and decreasing to 0.4 second at higher audio frequencies. What is wrong with that? This is quite a reasonable range for the personal project studio. The average bedroom is smaller in size than the average living room and would have a somewhat lower reverberation time, but still within a usable range. So, as for home acoustics, the reverberation time is probably acceptable.

Home acoustical noise

As for background noise, the usual home and neighborhood has a far higher noise level than one would like to have on a demo. About the only practical way around this is to record at 2 A.M. when the world is quiet. The problem then shifts to the tolerance of family and neighbors. This background noise problem has no easy solution, although working with a closer microphone will help. Perhaps the too-high background noise level on the demo will not be noticed because of the remarkable musical performance.

Home acoustics—modes

All small rooms, including living rooms and bedrooms, have modal resonances. At low audio frequencies, the wavelength of sound is of the same order as the dimensions of the room. This means that sound pressure peaks and nulls will shift around as the music shifts. The theory is covered in chapter 12, but with a predetermined space, the best approach is not to worry about modes in advance but rather to depend on a sound treatment that will control them.

Home acoustics—diffusion

In the recent past, diffusion was basically unavailable except for a small amount scrounged from surface irregularities, polycylindricals, and other geometrical approaches. This important characteristic of a sound field came within reach of even project studios through the discovery of the diffusion of sound by a series of parallel wells related to number theory by Manfred R. Schroeder (Schroeder 1975).

The idea was first put into practical, commercial form by Peter D'Antonio (D'Antonio 1984), whose diffusors are in practically every top-flight recording studio in the world. Turning his attention to the small project studio and its need for effective and inexpensive diffusors and other studio treatment units, his company, RPG Diffusor Systems, Inc., now offers the Abflector (Fig. 4-1), the B.A.S.S. Trap (Fig. 4-2), and the Skyline (Fig. 4-3).

The Abflector

The concept of a reflection-free zone was published in 1984 (D'Antonio 1984). This is a zone around the mix position in a control room or a listening position in a listening room. In both cases, the person is listening to a stereo pair of loudspeakers. The desirable direct sound from the loudspeakers is degraded by early reflections from walls, the ceiling, the console face, etc. These reflections are replicas of the program sound, but they arrive at

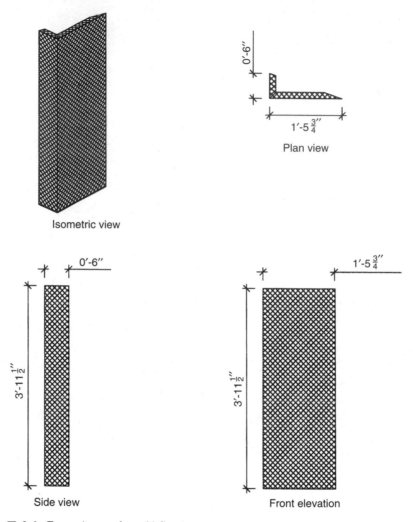

0'-6"

Plan view

1'-5 3/4"

Isometric view

0'-6"

3'-11 1/2"

Side view

1'-5 3/4"

3'-11 1/2"

Front elevation

■ **4-1** *Four views of an Abflector.*

the listener at slightly different times, creating comb filters that distort the direct sound. Eliminating or attenuating these early reflections greatly improves the sharpness of the stereo image, the breadth of the sound stage, and the general quality of the sound. In the average hi-fi listening situation these early reflections are perceived along with the direct sound, resulting in distortion. It might be called "ambience" and it might even be accepted and enjoyed, but it is distortion.

The Abflector of Fig. 4-1 is a splayed, absorbent, fabric-covered panel that can be mounted on walls and ceiling between the loudspeakers and the mix/listening position. At mid- and high-frequencies the Abflector deflects the energy of the early reflections toward the rear of the room. They serve another function, absorbing broadband sound; this is optimized by the spacing from the wall. Each weighs about five pounds, and can be mounted by adhesive or fasteners.

The B.A.S.S. Trap

Small rooms are generally subject to low-frequency reverberation problems resulting from the modal resonances of the room itself. This "boomy" sound in small rooms is well known. Because of the long wavelength of low-frequency sound, effective absorbers require much space. It is smart to utilize the corners of a room because it is space often unused. More importantly, all modes terminate in corners, making this location most effective for bass

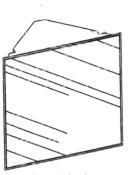

Isometric view

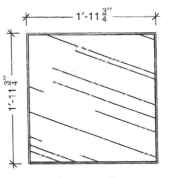

1'-11 3/4"

1'-11 3/4"

Front elevation

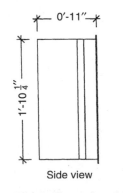

0'-11"

1'-10 1/4"

Side view

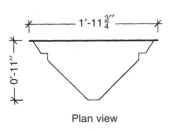

1'-11 3/4"

0'-11"

Plan view

■ 4-2 *Four views of a B.A.S.S. Trap.*

absorption. The B.A.S.S. Trap absorbers are trapezoidal units that are designed for corner mounting, Fig. 4-2. These units utilize a membrane with high internal loss and provide an absorption coefficient of 0.8 at 80 Hz as shown in Fig. 5-11. Each unit weighs 11 pounds, and can be mounted by adhesive or fasteners.

The Skyline

The Skyline (Fig. 4-3) is a very efficient omnidirectional diffusor needed to disperse first-order and other reflections from the rear wall and to contribute to achieving a diffuse field in the room. It is based on a primitive-root number sequence and the highest prime number ever used in diffusors. It is molded of fire-resistant poly-

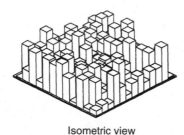

Isometric view

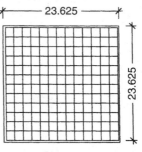

Plan view

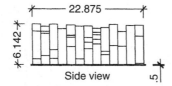

Side view

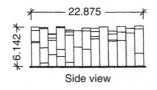

Side view

Side view

Side view

■ **4-3** *Four views of a Skyline.*

mer. Each unit weighs four pounds, and is mounted by adhesive or fasteners.

Step #1—getting started

In the studio design of Fig. 4-4 only two Abflectors are applied to each side wall. With this limited area, all of the early reflections might not be intercepted. Reflections from the front wall between the loudspeakers are untreated, as are reflections from floor and ceiling. Clearly, the space of Fig. 4-4 will be troubled by early reflections, because only the side walls are partially treated. There

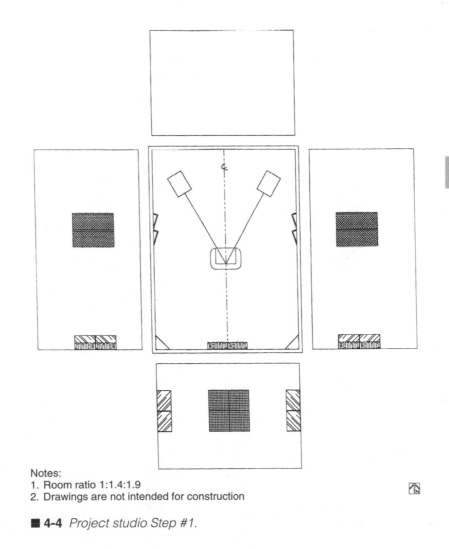

Notes:
1. Room ratio 1:1.4:1.9
2. Drawings are not intended for construction

■ **4-4** *Project studio Step #1.*

will be some improvement in sound quality because many of the side-wall reflections are intercepted by the dual Abflectors.

In Fig. 4-4, only two bass units are mounted in each of the two rear corners. The boominess will be reduced, but not eliminated. The four Skyline units on the rear wall intercept only a small fraction of the sound energy falling on the rear wall. These four will not provide a truly diffuse field, but their presence will be noted in a modest increase in sharpness of the stereo image, the depth of sound stage, and a feeling of being enveloped by the sound.

To keep costs down, Fig. 4-4 represents the very minimum defense against the effects of early reflections, the minimum of bass absorption, and the minimum of rear-wall diffusion. Much potential improvement remains.

Step #2—an intermediate solution

Using the studio acoustical treatment of Step #1 (Fig. 4-4) is bound to excite the user to want more of what has been offered in a limited way. For this, Step #2 treatment shown in Fig. 4-5 is presented as the next logical progression. The deficiencies of treatment in Step #1 are only in degree. More of same is the secret of Step #2.

There are now four bass absorbers in each of the two rear corners. If the ceiling height were limited to eight feet, this would be the maximum possible number in a corner. Incidentally, the first-order vertical mode with an eight-foot ceiling is 71 Hz. The absorption of the B.A.S.S. Trap is 100% at this frequency, as shown in Fig. 5-11. There are two other corners that could be filled with bass absorbers.

The number of Abflector units on each side wall are the same in both Step #1 and Step #2 designs, but there are now four Abflectors on the ceiling. The floor reflections could be reduced materially with a 6' × 6' rug.

Step #3—complete solution

The final design of Fig. 4-6 increases all units to their required level. Four of the B.A.S.S. Trap absorbers are now in each of the four corners of the room. Three Abflectors are now on each side wall, six on the ceiling and four on the front wall between the loudspeakers. Eight of the Skyline diffusors are now mounted on the

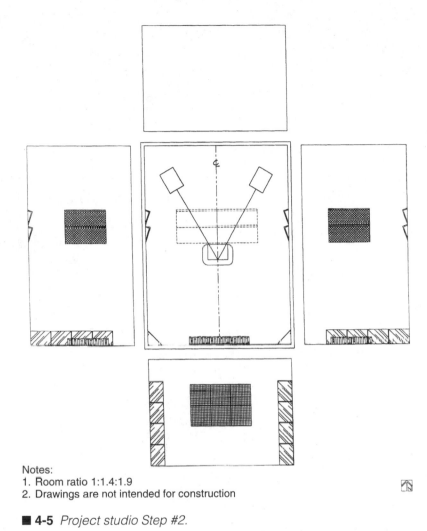

Notes:
1. Room ratio 1:1.4:1.9
2. Drawings are not intended for construction

■ **4-5** *Project studio Step #2.*

rear wall. With the present level of sensitivity, all of the potentially troublesome acoustical problems are now eliminated, and a mix done under these conditions should be transferable to other listening environments.

How about recording in this room?

The designs of Figs. 4-4, 4-5, and 4-6 are for listening or mixing with loudspeakers as the source of sound in all cases. What if the source of sound is a musician and it is desired to record him/her? Where is the musician to be positioned? Where should the microphone be placed? If the musician(s) are placed between the loud-

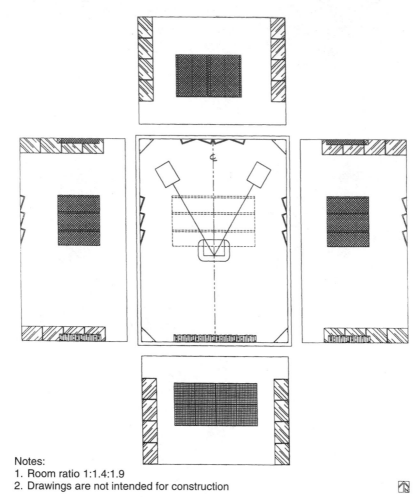

Notes:
1. Room ratio 1:1.4:1.9
2. Drawings are not intended for construction

■ **4-6** *Project studio Step #3.*

speakers and the microphone is placed at the sweet spot the recording would be quite free of early reflections. The side-wall, ceiling, and floor absorbers were placed by aiming on tweeters. With a broader source, such as one or more musicians located between the two tweeters, the early sound would not be intercepted completely by these absorbers, but it would be, partially.

But is it desirable to record musicians without early reflections? Eliminating the early reflections from wall, ceiling, and floor of the listening/mix room has been for the purpose of keeping the music from the loudspeakers free of them; of handling the loudspeaker sound on a completely neutral basis. The chances are good that music coming over the loudspeakers has its own burden of early reflections as part of what is called "good, high-quality" sound.

Therefore, any effort to record without early reflections could be considered specious; i.e., having a false look of genuineness. We have a good experimental recording venue in Fig. 4-6 by pushing the overstuffed chair to one side and using the main space of the room for the musicians. The absorbers on side walls, front wall, and ceiling will be in the direction of giving a good reverberatory condition. The absorbers in the corners will give what is called a "tight" bass sound. The large panel of diffusors will be greatly appreciated by the musicians. In short, all the elements are present for a good recording studio, even though they have all been placed with listening in mind.

The studio of Fig. 4-6 (or, to a degree, its early stages of Figs. 4-4 and 4-5) can be considered a good mix/listening room as well as a good home project recording studio. The acoustical elements that have been added are quite basic in their function, and are highly adaptable to space available in a home.

5

Small announce booth

THE SIMPLEST TASKS CAN TURN OUT TO BE THE MOST complex. The simplest and most widely used studio turns out to be an acoustical monstrosity. The term *announce booth* is a term probably from early radio broadcast days, when it was used for on-air identification announcements. The "booth" part is undoubtedly because of its comparative size.

Today, these small "announce booths" or "talk booths" are found in most radio and TV broadcasting establishments. The roving reporter and companion sound/camera operator cover the outdoor events, recording the reporter's voice on a portable tape recorder. Returning to the studio, the producer calls for additional narration by the reporter to cover scenes in which the reporter does not appear. These "post" segments of the reporter's voice are recorded in the talk booth, which might be a small closet under a stairway with some acoustic tile, carpet, foam, and/or egg cartons as acoustic treatment. The producer complains that the sound in live and post scenes do not match. "It sounds like one reporter outdoors, another on the post scenes. . . . make it sound like it's the same person!" That is the crux of the problem.

As a test to further illustrate the problem, record a voice outdoors, then with identical equipment record the same voice in a bathroom or shower stall. It is easy to hear the difference; it is difficult or impossible to make the two sound the same. Equalization will rarely be enough to match the two recordings.

The small-room problem

The sound field in a small room is dominated by resonances, which are completely absent outdoors. To minimize the variables in the comparisons to follow, a room having the dimensions of $6' \times 8'$ with an 8' ceiling height will be used for each of the three examples of this chapter. The lowest resonance frequency of this room is the axial mode associated with the longest dimension, the 8′ length of the room. Already another problem appears: the height is also 8′, which means that the effect of this lowest resonance frequency will be doubly strong. Table 5-1 lists all the axial modes of this 6′ × 8′ × 8′ room up to 300 Hz, beyond which modal colorations of the sound are minimum. Coincidences appear at 71, 141, 212 Hz, and a triple coincidence (triplidence?) at 282 Hz, which is a high-enough frequency to lead us to hope for a minimum coloration problem. Although the average difference between adjacent axial modal frequencies is 25 Hz, there is nothing average about the difference column in Table 5-1, which will result in a very irregular room response below 300 Hz. This means that extra absorption must be provided to control these axial modes. Rooms smaller than 6′ × 8′ × 8′ have even greater problems, as the axial modal frequencies are higher, but still able to degrade sound quality.

■ Table 5-1 Axial-mode study (room: 6′-0″ × 8′-0″ × 8′-0″)

	Length L = 8.0′ $f_1 = 565/L$	Width W = 6.0′ $f_1 = 565/W$	Height H = 8.0′ $f_1 = 565/H$	Arranged in ascending order	Diff.
f_1	70.6 Hz	94.2 Hz	70.6 Hz	70.6 Hz	0 Hz
f_2	141.3	188.3	141.3	70.6	23.6
f_3	211.9	282.5	211.9	94.2	47.1
f_4	282.5	376.7	282.5	141.3	0.
f_5	353.1		353.1	141.3	47.0
				188.3	23.6
				211.9	0.
				211.9	70.6
				282.5	0.
				282.5	0.
				282.5	70.6
				353.1	
					Average 25.7

Example 1—traditional talk booth

Example 1 is only to simulate the traditional talk booth and to discuss its problems. It is NOT a studio to be built. Of course, existing traditional booths are very different, and presenting one to represent them all is a fragile fiction. There is reason, however, to believe that acoustic tile and carpet are common to a great many existing talk booths, and these are featured in the following example.

Example 1—traditional talk booth

This assumed traditional talk booth is illustrated in Fig. 5-1. It is a small 6′ × 8′ × 8′ rectangular space of common 2 × 4 framing covered on both sides with 1/2″ gypsum board. With all its deficiencies, this structure should provide adequate insulation from outside noise except in extreme cases. The ceiling is covered with acoustic tile, as well as the walls above the wainscot strip. A heavy carpet with 40 oz pad covers the floor and the carpet (without the pad) is run up the wall to the wainscot strip. What kind of performance can be expected from this room?

Axial modes

To simplify the normal mode considerations of this small room, only the axial modes of the first order are included in Table 5-1. Table 5-1 tabulates all the frequencies of the axial modes of a 6′ × 8′ × 8′ space below 300 Hz. The frequency of the length axial mode (1,0,0) is the lowest, at 71 Hz. There will be no resonant support for sound of lower frequency than this. The length and the height of the room are the same at 8′. This results in coincidences and double potency at 71 Hz and at every multiple of 71 Hz. In the difference column there are coincidences at 71, 141, 212, and a triple coincidence at 282 Hz. Voice colorations must be expected at all these frequencies, especially the lower ones. Such coincidences could have been avoided (or at least minimized) by choosing more favorable room proportions.

In Fig. 5-2 the sound pressure distribution of the axial modes are described. The three little pressure graphs are moved outside the space to avoid confusion. The left graph is for the resonance set up between the east and west walls (0,1,0 mode). When this resonance is excited, the sound pressure over the entire east and west wall surfaces is high, with a null plane (a,b,c,d,a) extending from floor to ceiling.

The right graph of Fig. 5-2 is for the lengthwise mode (1,0,0). When this resonance is excited, the sound pressure is maximum over the entire surfaces of the north and south walls. There is a null plane e,f,g,h,e bisecting the room.

The graph in the upper right of Fig. 5-2 is for the axial mode set up at resonance between the floor and the ceiling (0,0,1). When this resonance is excited, sound pressure is high over the entire floor and ceiling surface with a null plane i,j,k,l,i bisecting the room.

These pressure maxima and minima dominate the acoustics of the space. The narrator could, for instance, very well sit with his head

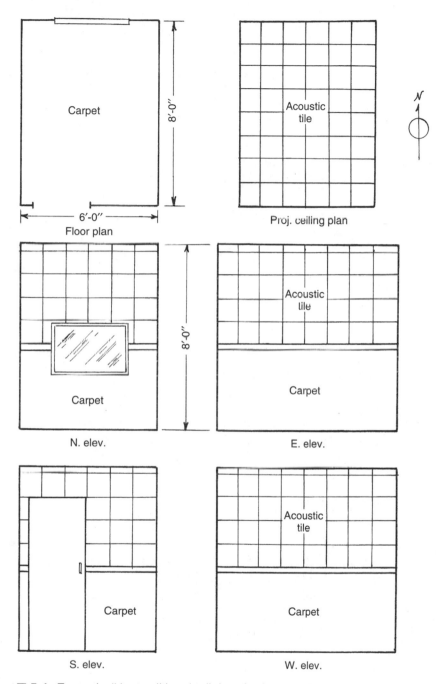

■ 5-1 *Example #1—traditional talk booth plan.*

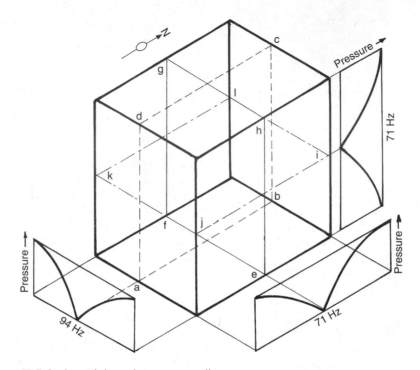

■ **5-2** *An axial-mode pressure diagram.*

positioned in or near all three null planes. The response of the microphone could fluctuate wildly as it is moved about in search of the best-sounding position. What the narrator hears and the microphone "hears" may be quite different.

Reverberation time

With the usual disclaimer on the application of reverberation time to a small room that has no reverberant field, it is still helpful in studying the distribution of absorbence present in the room. Table 5-2 shows the calculations applying to the carpet and acoustic tile. This results in exceptionally long reverberation time at 125 Hz, as shown in Fig. 5-3.

A "hidden" component of absorption of sound in the room is that diaphragmatic absorption of the gypsum board panels and the wood floor which is often overlooked. The relatively large areas of wall panels and floor mean that their resonance points and peaks of absorption would be at low frequencies. Absorption at low frequencies is just what is needed to control the axial modes, and to minimize the great variations of sound pressure in the room

■ Table 5-2 Reverberation time calculations (Voice booth—traditional: Example #1)

Material	S = area ft²	125 Hz		250 Hz		500 Hz		1 kHz		2 kHz		4 kHz	
		a	Sa	a	Sa	a	Sa	a	Sa	a	Sa	a	Sa
Carpet, heavy 40 oz pad	140	0.08	11.2	0.24	33.6	0.57	79.8	0.69	96.6	0.71	99.4	0.73	102.2
Acoustic tile ½"	155	0.07	10.9	0.21	32.6	0.66	102.3	0.75	116.3	0.62	96.1	0.49	76.0
(Total sabins)			22.1		66.2		132.1		212.9		195.5		178.2
(Reverb. time)			(0.85)		(0.28)		(0.10)		(0.09)		(0.10)		(0.11)
Drywall	224	0.29	65.0	0.10	22.4	0.05	11.2	0.04	9.0	0.07	15.7	0.09	20.2
Floor, wood	48	0.15	7.2	0.11	5.3	0.10	4.8	0.07	3.4	0.06	2.9	0.07	3.4
Total sabins carpet + tile + drywall + floor			94.3		93.9		198.1		225.3		214.1		201.8
(Reverb. time)			(0.20)		(0.20)		(0.10)		(0.08)		(0.09)		(0.09)

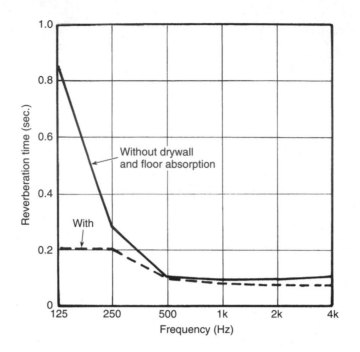

■ **5-3** *A reverberation time graph; Example #1.*

caused by the axial modes (Fig. 5-2). With the diaphragmatic absorption effective in the 70-250-Hz region, the modal maxima would tend to be brought down and the nulls filled in. The gypsum board and wood floor provide absorption in this needy region.

By adding the absorption due to the walls and the floor to that of the carpet and acoustic tile, the reverberation time is lowered materially as shown in Fig. 5-3. It is possible that in some cases a small room has been made usable by this hidden low-frequency absorbence although never figured into the calculations. Without depending too much on the precision of the calculated reverberation time, the 0.1 second of Fig. 5-3 is very low. A deadness would characterize this room, possibly too much for comfort and effectiveness.

Summary: Example 1

What problems would characterize the traditional talk booth of Example 1?

1. The axial modes would produce an irregular sound field. The diaphragmatic absorption of walls and floor might be insufficient to control the modes.

2. Excessive mid- to high-frequency absorption makes the room too dead.

3. Diffusion is needed to thoroughly mix the sound.

Example 2—voice booth with Tube Traps

This example of a voice booth is based upon the products of Acoustic Sciences, Inc. (Ref. ASC). Figure 5-4 shows 16″ quarter-round Tube Traps in the four corners and 9″ half-round Tube Traps on four walls and the ceiling. These alone constitute the acoustic treatment of the room. The half-rounds on walls and ceiling provide absorption and, in conjunction with the strips of reflective wall surface between the Tube Traps, diffusion of sound.

The perspective drawing of the talk booth (with two walls removed, Fig. 5-5), might help in understanding the conventional drawings of Fig. 5-4.

The half-round Tube Trap is constructed as shown in Fig. 5-6. It is a rigid, easily-handled unit with fabric cover. The sound absorption characteristics of the various forms of the Tube Trap are shown in Fig. 5-7. Note that in this figure the absorption per linear foot is given directly in sabins, or absorption units, rather than the usual area and absorption coefficient approach.

In the usual studio, early reflections from bare areas of floor, walls and ceiling are dominant and all the source of problems. These interact with each other, forming comb filter coloration of the sound. In this Example 2 there are many, many discrete reflections from the strips of wall between the Tube Traps. In fact, there are so many of these reflections, offset from each other by small increments of time, that the cloud of comb filters produced are not audible as colorations but as a pleasant ambience.

Example 2: Techron TEF measurements

Time-energy-frequency measurements were made by Acoustic Sciences Corporation in the small room of Fig. 5-8(A) which is constructed on the same principles as the one of Fig. 5-4.

This gives a truly microscopic view of the acoustical functioning of the room. In Fig. 5-8(B) the time scale runs from 0 to 80 ms. The spike on the extreme left is the arrival of the direct sound at the microphone. The uniform decay reveals a reverberation time of

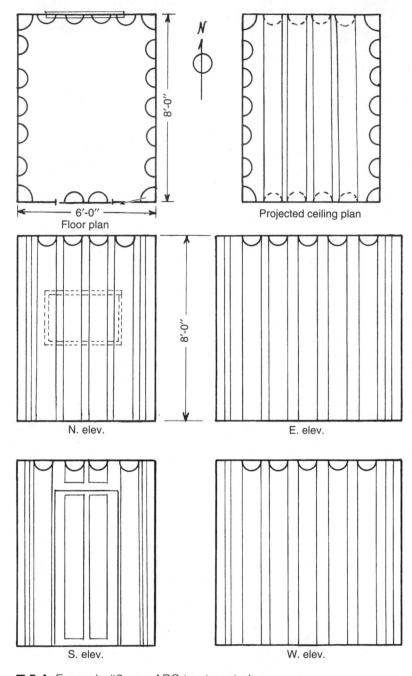

8'-0"

6'-0"
Floor plan

Projected ceiling plan

N. elev.

E. elev.

8'-0"

S. elev.

W. elev.

■ **5-4** *Example #2—an ARC treatment plan.*

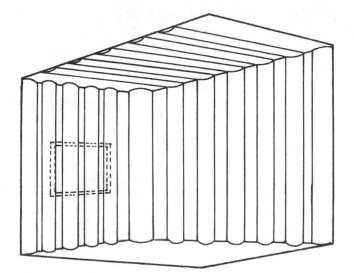

■ **5-5** *A perspective sketch; Example #2.*

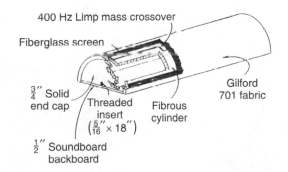

400 Hz Limp mass crossover

Fiberglass screen

$\frac{3}{4}''$ Solid end cap

Threaded insert $\left(\frac{5}{16}'' \times 18''\right)$

Fibrous cylinder

Gilford 701 fabric

$\frac{1}{2}''$ Soundboard backboard

■ **5-6** *Tube Trap construction details.*

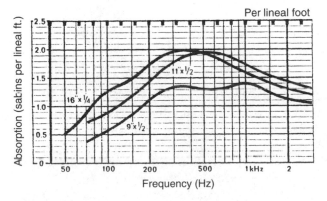

Per lineal foot

Absorption (sabins per lineal ft.)

2.5
2.0
1.5
1.0
0.5
0

11″x½

16″x¼

9″x½

50   100   200   500   1kHz   2

Frequency (Hz)

■ **5-7** *Tube Trap absorption.*

Example 2: Techron TEF measurements

about 0.08 second (80 ms). Figure 5-8(C) is an expanded view of Fig. 5-8(B), with time running from 0 to 20 ms. There is a 3-ms early arrival time gap between the arrival of the direct sound and the arrival of the stream of reflections from the Tube Trap grids on walls and ceiling. The "waterfall" of Fig. 5-8(D) reveals a very smooth, consistent, and quite diffused decay throughout the audible band.

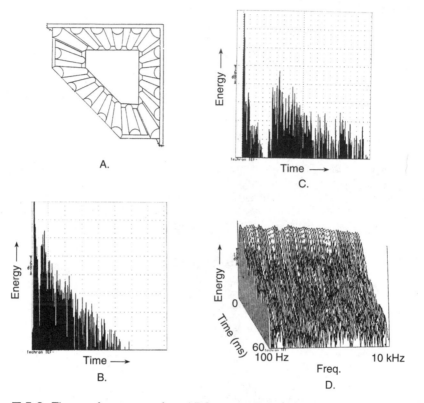

■ **5-8** *The performance of an ARC-type talk booth.*

Example 2—Reverberation time

The reverberation time calculations for the Tube Trap voice booth are shown in Table 5-3 and the results are plotted in Fig. 5-9. This very low reverberation time of about 70 ms compares with the TEF decay of Fig. 5-8(B). A similarly low reverberation time was calculated for the traditional booth of Fig. 5-3, which was overloaded with carpet and acoustic tile. The difference is that the dead characteristics of the Tube-Trap room come with highly diffused sound.

■ Table 5-3 Reverberation time calculations (Voice booth with ASC Tube Traps: Example #2)

Description	Length, ft.	125 Hz		250 Hz		500 Hz		1 kHz		2 kHz		4 kHz	
		abs/ft	abs	abs/ft	abs	abs/ft	abs	abs/ft	abs	abs/ft	abs	abs/ft	abs
9″ Tube Traps half-round	156	0.8	124.8	1.3	202.8	1.3	202.8	1.4	218.4	1.1	171.6	1.0	156.0
16″ Tube Traps quarter-round	32	1.4	44.8	1.9	60.8	1.9	60.8	1.6	51.2	1.3	41.6	1.1	35.2
Area sq ft, S		a	Sa	a	Sa	a	Sa	a	Sa	a	Sa	a	Sa
Drywall	224	0.29	65.0	0.10	22.4	0.05	11.2	0.04	9.0	0.07	15.7	0.09	20.2
Floor	48	0.15	7.2	0.11	5.3	0.10	4.8	0.07	3.4	0.06	2.9	0.07	3.4
Total sabins			241.8		291.3		279.6		282.0		231.8		214.8
Reverb. time =			(0.078)		(0.065)		(0.067)		(0.067)		(0.081)		(0.088)

$$\frac{(0.049)\,(\text{volume})}{\text{Total sabins}} = \text{sec.}$$

abs = absorption

Example 2—reverberation time

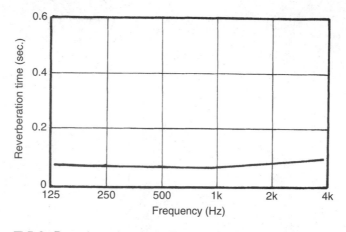

■ **5-9** *Reverberation time; Example #2.*

What are the human reactions within such a dead space? The narrator sitting in such a dead booth would have little acoustical feedback from the room. The narrator's own voice would sound somewhat unnatural to himself. If this is a problem, it can be rectified by providing the narrator with headphones and a quality playback of his/her own voice.

Example 3—voice booth with diffusors

The first voice booth, the not-so-good traditional kind, used absorption excessively. The second used ASC Tube Traps, and resulted in a new approach to the recording of quality voice signals from a small room. This example also produces satisfactory small-room conditions through the use of diffusing elements made by RPG Diffusing Systems, Inc.

The plan shown in Fig. 5-10 is based on the same 6′ × 8′ × 8′ room. The T-bar suspended ceiling is filled with twelve 2′ × 2′ Formedffusor quadratic-residue units as described in chapter 12.

These are not only diffusing elements; in this drop-ceiling mounting, they offer good mid- to low-frequency absorption as well. To control axial modes, two RPG B.A.S.S. (Bass Absorbing Soffit System) Traps are mounted in each corner. These contain a membrane with high internal losses mounted in a trapezoidal plastic cavity that fits into 90-degree corners. Preliminary absorption measurements on this unit are shown in Fig. 5-11.

Excellent absorption is provided by these B.A.S.S. Traps at the dominant axial modes of the room at 71 and 94 Hz. As the 8-foot

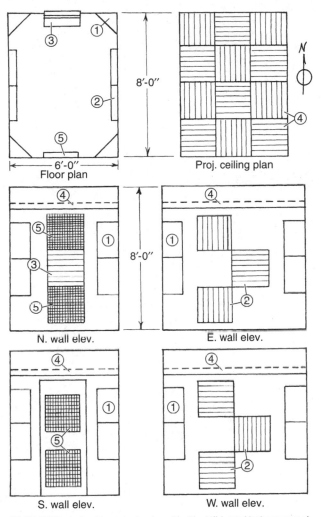

8'-0"

⑤

⑥'-0"
Floor plan

Proj. ceiling plan

N. wall elev.

E. wall elev.

S. wall elev.

W. wall elev.

(1) B.A.S.S Traps, low-frequency absorbers, 2′ × 2′ × 11″, located in the corners of the room, quantity eight. (2) Abffusors, 2′ × 2′ × 4″, absorber and diffusor of sound, six each on east and west walls, two on north wall. (3) Diviewsor, a diffusor with plexiglass panels making it transparent, 2′ × 2′ × 9″. Placed over the observation window. (4) Formedffusor, mounted in standard T-bar suspended ceiling, 12 required, 2′ × 2′ × 4″. These accomplish both diffusion and absorption. (5) Skyline, 2-dimensional diffusing element located on door, 2′ × 2′ × 6″, two required.

■ **5-10** *Example #3—an RPG treatment plan.*

ceiling height is widely used in smaller studios, the 71 Hz normal mode associated with that dimension receives maximum attenuation. The B.A.S.S Trap units are mounted in the corners where modal pressures are highest.

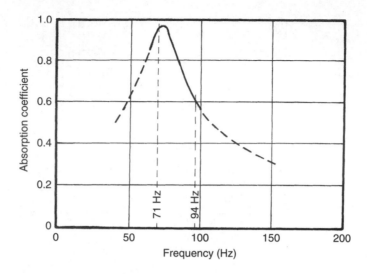

■ **5-11** *Absorption of a B.A.S.S. Trap.*

Abffusor units are mounted on the east and west walls, three to a side. These are positioned to minimize chance of flutter echo and to distribute the diffusion effect. These units have an excellent wide-band absorption characteristic, as shown in Fig. 5-12.

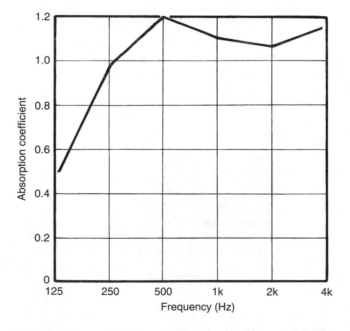

■ **5-12** *Absorption of an Abffusor.*

Two 2′ × 2′ Skyline units are mounted on the door in the south wall. Two others are mounted above and below the observation window in the north wall. These have maximum diffusion in two dimensions with a minimum of absorption. The dramatic appearance of these diffusors is shown in chapter 12 and Fig. 3-11.

The only remaining diffusor is mounted over the observation window of the north wall. This 2′ × 2′ Diviewsor is a quadratic-residue diffusor based on the prime 7 sequence. It is a conventional type-734 diffusor (see Fig. 12-15), except that the panels are of transparent Plexiglas. This allows visual communication through the glass window.

Example 3—reverberation time

A natural question at this point is, "Example 3 has plenty of diffusion, but is there enough absorption to keep it from being too 'bright'?" A few calculations will answer this question. Table 5-4 shows the reverberation time calculations for the voice booth with RPG diffusors. No sound absorption material has been added to the room of Example 3 apart from that intrinsic to the diffusing elements and the structure. Both the Abffusors and the Formedffusor have excellent absorption in addition to their main function as diffusors. From the calculations of Table 5-4 come the reverberation time data plotted in Fig. 5-13.

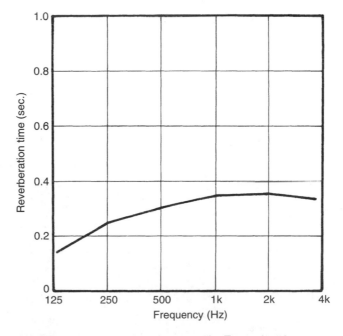

■ **5-13** *A reverberation time graph; Example #3.*

Example 3—reverberation time

■ Table 5-4 Reverberation time calculations (Voice booth with RPG diffusors: Example #3)

Description	Area sq ft, S	125 Hz		250 Hz		500 Hz		1 kHz		2 kHz		4 kHz	
		a	Sa	a	Sa	a	Sa	a	Sa	a	Sa	a	Sa
B.A.S.S. (8)	32	0.4	12.8	0.3	9.6	0.1	3.2	0.05	1.6	0.03	1.0	0.02	0.6
Abffusors (6)	24	0.82	19.7	0.90	21.6	1.07	25.7	1.04	25.0	1.05	25.2	1.04	25.0
Formedffusors (12)	48	0.53	25.4	0.37	17.8	0.38	18.2	0.32	15.4	0.15	7.2	0.18	8.6
Drywall	224	0.29	65.0	0.10	22.4	0.05	11.2	0.04	9.0	0.07	15.7	0.09	20.2
Floor	48	0.15	7.2	0.11	5.3	0.10	4.8	0.07	3.4	0.06	2.9	0.07	3.4
Total sabins			130.1		76.7		63.1		54.4		52.0		57.8
Reverb. time, sec. $= \dfrac{0.049 \times \text{volume}}{\text{Total } Sa}$			(0.014)		(0.25)		(0.30)		(0.35)		(0.36)		(0.33)

Small announce booth

■ Table 5-5

Example	Acoustic Treatment	Normal Mode Treatment	Reverberation Time at 500 Hz	Mic. Placement Sensitivity	Narrator Hearing Own Voice
#1	Ac. tile, carpet	* (only)	100 ms	Bad	Needs headphones
#2	ASC Tube Traps	* plus Tube Traps	70 ms	No problem	Needs headphones
#3	RPG diffusors	* plus diffusors	0.3 sec	No problem	Natural

* Drywall and floor diaphragmatic absorption.

Example 3—reverberation time

The reverberation time of approximately 0.3 second should be about right for such a small studio. It is suggested that anyone building a small studio like this and treating it with diffusing elements alone should listen analytically with the floor bare, and then add a 5′ × 7′ rug or even carpet the entire floor to bring the reverberation time down to about 0.2 second.

Example 3—evaluation

With so many diffusing units, the sound field in the room should be thoroughly diffused. The narrator should experience a sense of being in a larger space, and the sound he hears should have a superior fullness and clarity. The microphone pickup should be relatively independent of microphone location. The narrator's perception of his own voice should be quite natural.

Comparison of examples 1, 2, and 3

The same too-small studio (6′ × 8′ × 8′) has been treated in the traditional way and two ultra-modern ways using equipment that has been available only within the last decade or so. The results are summarized in Table 5-5.

Control rooms

6

THE MIXER IN THE CONTROL ROOM IS RESPONSIBLE FOR evaluating sounds coming from the monitor loudspeakers, whether the source is live music from the studio, from an existing recording, or is the sound of many different tracks of a current project being produced. The mixer must judge sound quality, which involves frequency response irregularities, level changes, distortion, slight tonal differences, timing of segments, and scores of other minute details.

For his listening this mixer needs an environment that is absolutely neutral. The acoustical link between the monitor loudspeakers and the mixer's ears must not add any perceptual changes. That is the very important burden of the control room: to deliver the sound without adding anything to it or taking anything away. (How different from listening to the home hi-fi!) As the sound travels from the home loudspeakers to the listener's ear, much is added. We become accustomed to this added "texture" and manipulate it to our pleasure. In the control room, great effort is expended to eliminate this "texture," but using headphones might be the only way to eliminate all of it, and that is an unsatisfactory and unnatural solution.

Early reflections

The science of control room acoustical design has made great strides since Don Davis and Chips Davis experimented with making the front end of a control room completely absorbent and noting a great improvement in the quality of sound from the monitor loudspeakers (Davis 1980). The reason they applied all that glass fiber around the loudspeaker end of the control room was to reduce early reflections and their degrading effect on sound quality, and it worked.

Adding absorbent material to control room surfaces is one way of reducing the effect of early reflections, but it absorbs precious sound energy as well. The mixer compensates by increasing the gain to bring the sound level back up to his desired level. Modern control rooms minimize the use of absorbent, with the result that amplifiers and loudspeakers can run at a lower level (i.e., with lower distortion).

Combing of early reflections

The direct sound arriving at the mixer's ears is clean and pure as far as the acoustical path is concerned. It is the desired sound. Sounds reflected from side walls, ceiling, or floor arrive with appreciable levels and at slightly different times. When replica signals with small time differences between them combine, comb filters are created. This is true whether reflections combine with each other or reflections combine with the direct ray. The result of this combining is the changing of a flat response to a series of peaks and nulls down through the audible spectrum: distortion, in other words. In the home hi-fi situation one gets used to this distorted "edge" on the signal created by this combing, often called "ambience." In the control room, such distortion must be avoided at all costs.

Examples of combing

To pursue the question of "What is combing?," refer to Fig. 6-1. This is a highly simplified laboratory setup in which a sound source, a microphone, and a baffle (a piece of plywood) are arranged in that order in a straight line. In Fig. 6-1 the left peak is the direct sound picked up by the microphone on its way to the baffle. The second one is a specular reflection from the baffle. These two combine with a time difference (phase) between them

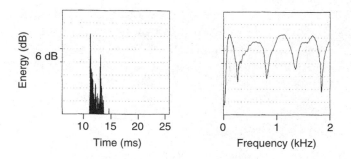

■ 6-1 *A comb filter example.*

creating the typical peak/null comb-filtered response. This distortion is normally quite audible, and therefore to be avoided.

Figure 6-2 is not a laboratory setup, but an actual case in a control room. In Fig. 6-2(A), the direct spike on the left and the specular reflection spike on the right are separated in time by about 7 ms. In combining, the resulting response is changed from a flat condition to a series of peaks and nulls. As previously stated, such a comb filter response results in audible distortion.

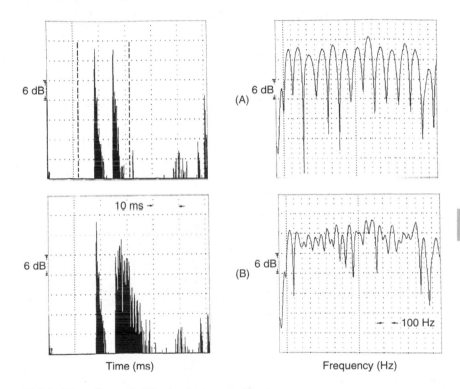

■ **6-2** *The effect of diffusion on comb filtering.*

In Fig. 6-2(B), the direct spike is the same as the one above, but the reflection (instead of being specularly reflected) has fallen on a diffusor, and has been diffused and spread over 12 ms of time. When the diffused reflection combines with the direct spike some comb filtering results, but the amplitudes of peaks and nulls are reduced and the fluctuations are of a higher frequency. From a perception standpoint, the response of (A) is very audible, but that of (B) is essentially inaudible. By diffusing early reflections before they combine with the direct or other reflections, the audible distortion is practically eliminated.

Reflection-free zone

A reflection-free zone about the mixer's position can be achieved by covering the offending reflecting surfaces with absorbing material. Too much absorbing material in a control room absorbs too much signal energy; however, some absorption is needed to adjust the reverberation time of the space. An approach to a reflection-free zone that does not rob the space of signal energy is that of shaping the reflecting surfaces so that reflections are directed away from the mixer position toward the rear of the room.

Figure 6-3 is an outline sketch of the front part of a control room. By splaying the side walls as shown, the early first-order reflections are directed toward the rear of the room instead of the mixing position as rectangular side walls would do. The early reflections are directed toward the rear of the room to be diffused and returned to the mixer as desired, providing helpful, reverberatory sound. The shaded area of Fig. 6-3 is a reflection-free zone of sufficient size to cover the console, the mixer, and the producer's position behind the mixer.

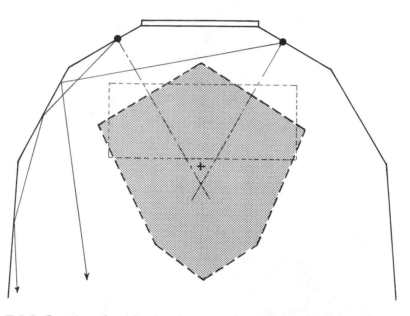

■ **6-3** *Creation of a reflection-free zone by splaying the side walls.*

A control room constructed to achieve this reflection-free zone and with proper diffusion at the rear wall is characterized by:

☐ Clear sound, free from combing distortion.
☐ An extremely wide "sweet spot."

☐ A spatial impression of being enveloped by the sound.

☐ Accurate perception of the stereo image.

☐ Flat low-frequency response with uniform modal decay.

☐ Reproducibility and transferability of product.

Two-shell control rooms

Control rooms are often built on the "two-shell" plan. Only a light-weight surface is necessary to deflect the mid- to high-frequency early sound in the proper direction. Therefore, the inner shell of a control room built to a shape required for control of early sound can be of light construction. Low-frequency energy, with its very long wavelengths, tends to permeate the space and must be dealt with independently. The outer shell is heavy, and is built to control the low-frequency normal modes and to shield the inner room from environmental noises. The outer shell is often constructed of concrete blocks, normally with voids filled with well-rodded concrete. The shape of the outer shell is somewhat controversial: some designers favor a trapezoidal shape, some a rectangular. The trapezoidal shape does nothing toward reducing modal problems, it only shifts peaks and nulls to indeterminant, asymmetrical positions. The rectangular outer shell, at least, yields predictable, symmetrical modal patterns.

Example 1—control room

A really good control room is a very expensive structure. If money is not available to do it right, compromises must be made. Several less-than-perfect (but still good) control rooms will be dissected as areas of compromise are studied. A rectangular control room located, no doubt, in an existing structure is shown in Fig. 6-4. To avoid soaking up too much signal energy, absorbing material is minimized in favor of diffusing elements. The loudspeakers (1) are aimed in the normal hi-fi configuration. The RPG Abffusor is both a diffusor and an absorber (see Fig. 5-12). Abffusor panels (2) are mounted on the ceiling between loudspeaker and listening position. Other absorbing/diffusing panels (3) of the same type are mounted on the left and right walls to intercept the reflections that would go toward the mixer.

The mixer's head is located between the two parallel wall surfaces. Flutter echoes between the left and right walls are a threat. These could be defeated with absorbent (which we wish to avoid). RPG

Example 1—control room

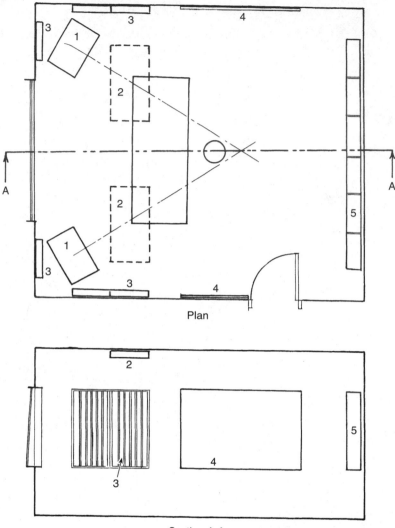

Plan

Section A-A

■ **6-4** *Control room example—rectangular walls.*

Flutterfree panels of surface-mounted hardwood molding are installed on both walls (see Fig. 12-20). These strips of high-frequency diffusor will eliminate any tendency toward flutter. The diffusor group at the rear of the control room is composed of six 2′ × 4′ RPG QRD 734s, half of them with vertical wells and half with horizontal wells so as to give diffusion in both the horizontal and vertical planes (see Fig. 12-15).

Example 2—control room

Example 2, shown in Fig. 6-5, controls early reflections by a combination of room shaping and absorption. It has a shape that would fit well into a rectangular, heavy outer shell. The loudspeaker's (1) center lines intersect close behind the mixer's head as usual. The side walls are splayed enough to take care of flutter echo. If this

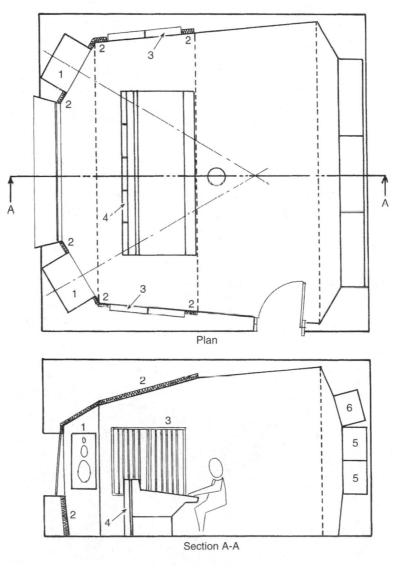

Plan

Section A-A

■ **6-5** *Control room example—slightly-splayed walls.*

Example 2—control room

side-wall splaying is not sufficient to cope with the early-reflection path, RPG Abffusors (3) are set into the side wall.

The ceiling, side, and front walls are covered with Absorbers (2), which are composed of a fabric-covered graduated-density glass fiber. This covering extends toward the rear as far as the armrest of the mixing console. The back of the console (4) and the wall under the window constitute a quasi-cavity, which can resonate at its own frequency. To control this resonance and to avoid "bass buildup," Absorbers (4) are affixed to the rear of the console. They work with the Absorbers mounted under the window.

The rear-wall diffusor array is made up of two rows of three QRD 1925 (5) diffusors set horizontally with their wells vertical (see

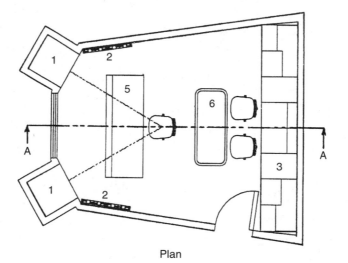

Plan

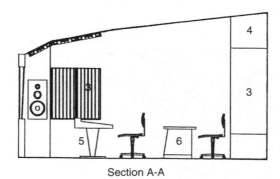

Section A-A

■ **6-6** *Control room example—partially splayed walls.*

Fig. 12-17). On top of them is a row of three QRD 725 (6) diffusors set horizontally with their wells horizontal. These last three (6) are angled forward. This combination gives good diffusion in the horizontal and the vertical plane.

Example 3—control room

The third control room example is shown in Fig. 6-6. Elements of similarity to the previous two are evident, with a few significant differences. This room features slightly splayed side walls at a single angle. The splaying is not enough to eliminate the need for side-wall absorption/diffusion, but it is enough to eliminate flutter between the side walls. Room for a producer's desk (6) has been found behind the mixer's position. The monitor loudspeakers are

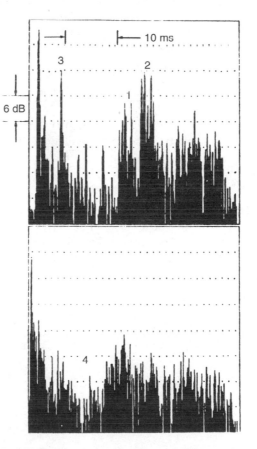

■ **6-7** *Echogram—the effect of diffusors in a control room.*

Example 3—control room

recessed into soffits and mounted flush. The Abffusors (2) are mounted on the side walls to intercept the first-order reflections aimed at the mixer. Others are also mounted on the underside of the ceiling to do the same thing.

At the rear of the room a large diffusor (3) is placed. No specifications for this rear diffusor are on the sketch. Three interesting Diffractal units are available from RPG Diffusor Systems, Inc. as follows:

Model Number	Size	Frequency Range
DFR-72MH	2′ × 4′	400 Hz–17 kHz
DFR-82LM	3′ × 11′	100 Hz–5 kHz
DFR-83LMH	3′ × 16′	100 Hz–17 kHz

Some of these are giants. Another possibility is making the low-frequency diffusor of ordinary concrete blocks, or of the special RPG DiffusorBlox. Mounting diffusors within diffusors (the diffractal principal) makes possible very wide-band operation. Those reaching down to 100 Hz certainly take care of potential modal problems in the room.

The effect of the rear-wall diffusor action is clearly shown in Fig. 6-7 (D'Antonio 1984). These records were taken in two very similar control rooms, the top one without rear diffusors and the bottom one with rear diffusors. In the top record, the early energy is confined primarily in the groups of specular reflections (1) and (2) arriving at about 17 and 21 ms. The reflection off the console (3) is also a very prominent threat. The lower record shows the effect of the rear-wall diffusor. The valley of low reflections (4) is the arrival time gap of about 17 ms width; after that come the highly diffuse reflections that add body and ambience to the direct signal.

7

Audio/video/film workroom

IT TAKES MORE THAN PERFECT WORKING CONDITIONS TO turn out the perfect job, but there is no denying that poor working conditions only add to the difficulty of turning out an acceptable job. This is true in life in general, and is even more true in audio, video, film production, and postproduction work.

Audio fidelity and near-field monitoring

Little loudspeakers perched on the instrument bridge of audio consoles seem destined to replace the large monitors. This is at least true in the minds of many operators of such consoles. Hearing the sound "close up" gives these operators a confidence they do not get out of the big monitors. Some justify the smaller loudspeakers on the assumption that they help appreciate how the program sounds to those out in consumer land. Others use them a great proportion of the time because they like the more honest sound of the small loudspeakers.

Sad to say, the near-field monitors often do give a more accurate rendition of the program material than the big, expensive loudspeakers. Why does this situation exist? It is not a matter of one loudspeaker system against another, it is a matter of early reflections adding distortion to what the operator hears from the large loudspeakers but avoids by listening "up close" on the near-field loudspeakers. These early reflections are separated in time because they travel slightly different distances. The different reflections are replicas of the same signal with time shifts between them.

When a signal is combined with a replica of itself shifted in time, the two signals pull in and out of phase down through the spectrum creating a comb filter. Many reflections combining form many comb filters, which result in distortion and coloration of the sound. The superior sound from the large monitor loudspeakers is distorted by the early reflections. Eliminating the reflections restores the superior sound at the listening position. Listening "up close" to the near-field loudspeakers discriminates against the reflections, but accepts the inferior sound quality of the smaller loudspeakers.

In the design of this production workspace, priority is given to eliminating early reflections at the workstation position so that there will be no need for near-field monitors. That is not as simple as it sounds; early reflections are easily generated, and require constant vigilance to avoid their deteriorating effect. The ever-growing influence of digital techniques is forcing more attention on sound quality—not just low noise, but a clear and more accurate sound as well.

Axial-mode considerations

The first step in minimizing the deleterious effect of axial modes is to select proper proportions for the room. This is discussed in chapter 12, especially Table 12-3. Sepmeyer's "C" ratio, $1.00 \times 1.50 \times 2.33$, is used in this room; it gives a volume of about 7000 ft^3 for the room of Fig. 7-1, excluding the space behind the loudspeakers. Table 7-1 lists the axial-mode frequencies for this room. Listing these frequencies in ascending order and noting the difference in frequency between adjacent modes gives a good appraisal of possible modal problems. The mean (average) difference is 9.5 Hz, although the real differences vary from 0 to 20.2 Hz. A near-coincidence occurs at 141 Hz and 282 Hz, and a definite coincidence at 235 Hz.

Only the 141-Hz mode threatens a sound coloration; the others are high enough in frequency to avoid coloration effects. The 141-Hz coincidence is associated with both the 12-foot ceiling height and the length of the room. The chances of the 141-Hz coloration being audible are rather slim, because the tangential and oblique modes (even though of lower potency) will tend to fill in the gaps between the axial modes. The beauty of a larger room like this is that problems of mode spacing tend to disappear.

Monitor loudspeakers and early sound

The general plan of the room is shown in Fig. 7-1. To give the operator at the workstation the best sound possible from the monitors, two loudspeaker soffits are centered on 60-degree lines converging slightly behind the workstation position. The function

■ Table 7-1 Axial-mode distribution (room: 12.0′ × 19.20′ × 27.96′)

	L H = 565/27.96	W W = 565/19.20	H H = 565/12.0	Ascending order	Diff.
f_1	20.2 Hz	29.4 Hz	47.1 Hz	20.2	9.2
f_2	40.4	58.9	94.2	29.4	11.0
f_3	60.6	88.3	141.3	40.4	6.7
f_4	80.8	117.7	188.3	47.1	11.8
f_5	101.0	147.1	235.4	58.9	1.7
f_6	121.2	176.6	282.5	60.6	20.2
f_7	141.5	206.0	329.6	80.8	7.5
f_8	161.7	235.4		88.3	5.9
f_9	181.9	264.8		94.2	6.8
f_{10}	202.1	294.3		101.0	16.7
f_{11}	222.3			117.7	3.5
f_{12}	242.5			121.2	20.1
f_{13}	262.7			141.3	0.2
f_{14}	282.9			141.5	5.6
f_{15}	303.1			147.1	14.6
				161.7	14.9
				176.6	5.3
				181.9	6.4
				188.3	13.8
				202.1	3.9
				206.0	19.3
				222.3	13.1
				235.4	0.0
				235.4	7.1
				242.5	20.2
				262.7	2.1
				264.8	17.7
				282.5	0.4
				282.9	11.4
				294.3	8.8
				303.1	
				Mean	9.5
				Std. Dev.	6.4

of the soffits is to allow the faces of the monitors to be made flush with the angled wall. The corners of a stand-alone loudspeaker cabinet radiate sound through a wide angle by diffraction from the corners. This diffracted sound from loudspeaker edges contributes to the early sound problem directly and by reflections from the wall behind the loudspeakers. By making the faces of the

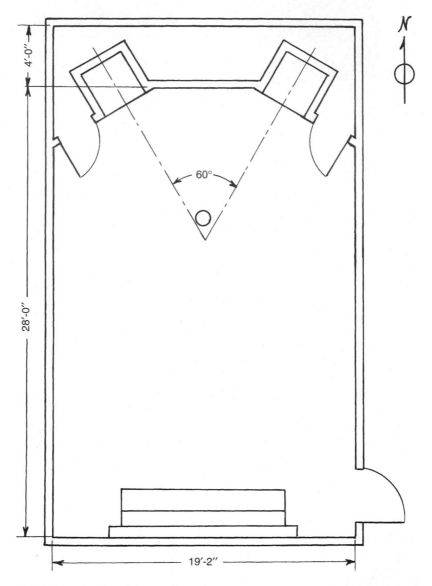

■ 7-1 *The AVF workroom plan view.*

monitor cabinets flush with the wall in which they are set, the diffraction effects and early sound from this source are eliminated.

A consideration of the good and bad direct rays from the monitors is shown in Fig. 7-2. The light lines indicate direct rays that bathe the operator in desirable sound. The broken, heavier lines are reflected from the walls, floor, and ceiling. Because they arrive at different times, comb filter distortion results. These can be controlled by locating areas of absorbent material at the indicated positions on the side walls and on the ceiling over the head of the operator (Fig. 7-8). Reflections from the floor are absorbed by the carpet.

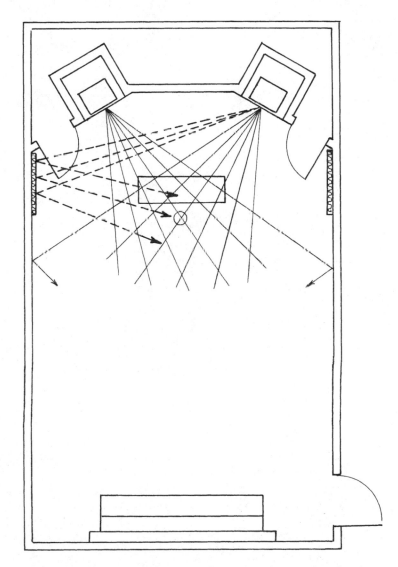

■ **7-2** *The AVF workroom plan—direct sound vs. early reflections.*

Late sound

Once the early sound problems are cleared up by the use of soffits and patches of absorbent, the operator (as mentioned) is bathed in direct sound from the monitors free from coloration-producing early reflections. This direct sound allows accurate perception of what is coming through the monitors, but it is only a part of the whole perception experience. The late sound made up of reflections from the rear of the room complete the perception. A rather detailed picture of both the early and the late sound is given in Fig. 7-3.

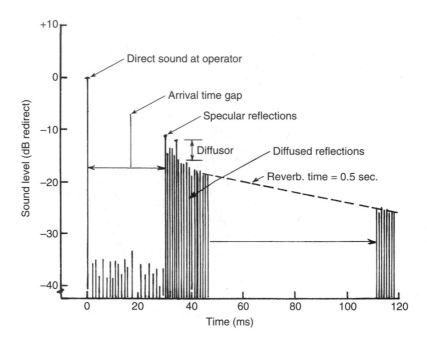

■ **7-3** *A reflection diagram for a workroom.*

The direct sound reaching the ears of the operator at the workstation establishes both the zero level and the zero time of this sketch. Following the arrival of the direct sound, a period of relative silence sets in as the sound passes the operator on its way to the rear wall. This is not a completely silent period; there are always minor reflections of low level. These miscellaneous reflections are 30 dB or so down, and hence less significant in the overall perception.

In Figs. 7-1 and 7-2 the distance to the workbench is about 17 feet, and the wall is about 20 feet from the operator. The direct sound must travel about nine feet from the loudspeakers to the operator, and must travel 34 to 40 feet more to get back to the operator as reflections from the face of the workbench and the wall. The level of the reflection from the workbench will be about 20 log 9/34 = −11.5 dB, and the level of the reflection from the wall will be about 20 log 9/40 = −12.9 dB. These rough estimates, based on inverse-square law propagation, are quite accurate enough for present purposes.

The delays of these two rear reflections are about 34 ft/1130 ft/sec or 30 ms, and 40 ft/1130 ft/sec = 35 ms. This information allows them to be plotted in Fig. 7-3; the reflection from the workbench is −11.5 dB at 30 ms of delay, and the reflection from the wall is −12.9 dB at 35 ms delay. These specular reflections are plotted on Fig. 7-3.

The rear-wall sound also falls on the bank of Skyline diffusors. The level of the return from the diffusors is 6 or 8 dB less than what falls on the diffusor face due to normal diffusor action. Actual measurements (see Fig. 12-11) reveal this. The flood of rear-wall reflections, both specular and diffused, decay at a rate determined by the reverberation time of the room. Our goal is a reverberation time of about 0.5 second, and that is the slope drawn in Fig. 7-3.

A substantial portion of the sound falling on the rear wall hits the reflection phase-grating diffusors and is diffused both in the horizontal and the vertical directions. The reflections from the workbench and the rear wall not covered with diffusors are mixed with the diffused sound and returned to the operator's ears delayed in time. The round-trip of the direct sound sweeping past the operator to the rear workbench and wall and returned to the operator's ears is such as to make the late returning sound 30 to 35 milliseconds later than the direct sound.

The arrival time gap in the room characteristic is very important to the operator, enabling him (during this gap of room "silence") to hear the arrival time gap in the program material.

Reverberation time

A room of 7000 ft^3 is a bit out of the "small room" category, and approaches the condition of thoroughly mixed sound. The concept of reverberation time can safely be applied in a room this size. Table 7-2 shows the calculations of reverberation time of the room.

Starting with no specific room treatment, the major absorber is the 1140 ft^2 of gypsum board surface with its diaphragmatic absorption. As we see the gypsum board peak in the low-frequency

■ Table 7-2 Reverberation calculations (Film/video/audio workroom)

Description	Area sq ft, S	125 Hz a	125 Hz Sa	250 Hz a	250 Hz Sa	500 Hz a	500 Hz Sa	1 kHz a	1 kHz Sa	2 kHz a	2 kHz Sa	4 kHz a	4 kHz Sa
Drywall	1148	0.29	332.9	0.10	114.8	0.05	57.4	0.04	45.9	0.07	80.4	0.09	103.3
Carpet, heavy, 40 oz pad	589	0.08	47.1	0.24	141.4	0.57	335.7	0.69	406.4	0.71	418.2	0.73	430.0
Wall & ceiling panels, 703, 2″ flat on wall	48	0.24	11.5	0.77	37.0	1.13	54.2	1.09	52.3	1.04	49.9	1.05	50.4
Total absorp. Sa			391.5		292.7		497.3		504.6		548.5		583.7
Rev. time = 0.049V			(0.88)		(1.18)		(0.77)		(0.69)		(0.63)		(0.59)
= [(0.049)(7068)]/Sa													
= 346.3/Sa													
= seconds													

region, it is natural to think of carpet (with its peak absorption in the higher frequencies) to compensate. This justifies carpet over the entire floor surface. The only other absorbing areas, and these are of minor effect because of their small size, are the two wall and one ceiling 4′ × 4′ panels to intercept the early reflections. The absorption of the gypsum board surfaces, the carpet, and the small panels is computed in Table 7-2. The resulting reverberation times for each frequency are plotted in Fig. 7-4.

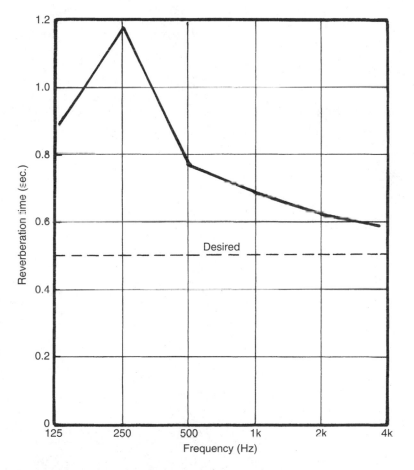

■ **7-4** *A reverberation time graph for a workroom.*

The gypsum board brings the 250-Hz reverberation time down from a very high value of 1.18 seconds, but it does not bring it down far enough. Compared to the desired reverberation time of 0.5 second, the values above 500 Hz are not very far off. About 400 ft^2 of Helmholtz resonators tuned to 250 Hz would bring this 1.18-seconds peak down to 0.5 second, but this would be unwise at this stage. There are too many uncertainties in gypsum board and carpet absorption to be that specific. A wiser approach would be to wait until the structure is built and the carpet laid, and then to measure the reverberation time to see precisely what correction is needed. The correction will be relatively small.

A goal of a 0.5-second reverberation time has been indicated, but the client might very well prefer 0.4 second (a slightly dryer acoustic) or 0.6 second (a slightly more lively acoustic). Such a preference and the measured values would provide the basis for the right correction.

Workbench

The south elevation of Fig. 7-5 shows the group of $2' \times 2'$ RPG Skyline diffusors. This will be a very dramatic panel made up of 18 of the units shown in Fig. 7-6. These units diffuse sound in both horizontal and vertical directions.

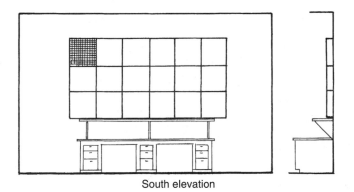

South elevation

■ **7-5** *South elevation, workroom.*

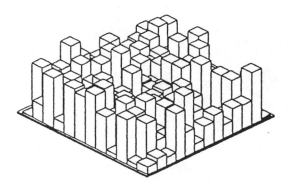

■ **7-6** *An isometric drawing of a Skyline diffusor.*

Beneath this diffusing area a workbench is suggested. Any reflections from this bench will simply add to the rear wall specular and diffuse reflection mixture. If placed anywhere else in the room, reflections from the bench would ruin the pristine purposefulness of Fig. 7-3 and fill in the important gap with spurious reflections.

The operator at the workstation will need a helper to do the many jobs to keep the workstation efficiently occupied. Certain equipment will be necessary to make this helper efficient. A shelf at the bottom of the diffusor panel is intended to hold such equipment. A few such pieces will have a negligible effect on the functioning of the diffusors.

Operator's workstation

Figure 7-7, the north elevation, shows the relationship of the operator to the monitor loudspeakers and the video screen. The line of sight is high enough that there is room for equipment without blocking the view, but care must be exercised in this regard.

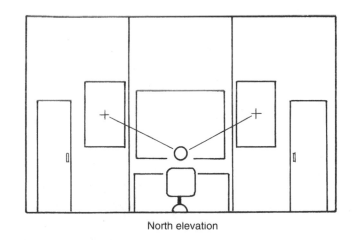

North elevation

■ **7-7** *North elevation, workroom.*

Racks of auxiliary equipment can be mounted under the desk on either side of the operator's feet. Another desk should not be placed behind the operator; equipment on it would send comb filtering reflections back to the operator's ears. Doors have been suggested to make the space behind the monitors available for storage, etc.

Figure 7-8 presents a side view of the operator's position in the west elevation view. The relative position of the side wall and ceiling early-sound panels is shown.

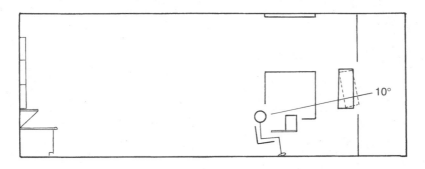

Should the monitor loudspeakers be vertical, or inclined downward to place the operator on the axis of the monitor? It would be best to be on this axis, but sound 10 degrees off-axis in high-quality monitors is practically the same. The simplest procedure is to accept a vertical partition and 10-degree off-axis sound. The partition could be made in two sections, one inclined at the same horizontal and the same vertical angle as the face of the left loudspeaker, and the other inclined to coincide similarly with the face of the right loudspeaker. This would cause some minor problems with the doors and the video screen which could be worked out. Monitor flushness must be considered a major issue in the battle against comb-filtering reflections (see chapter 6).

Lighting

It is suggested that the ceilings and upper walls be painted flat black, and that shaded light fixtures be hung from the ceiling. Track lighting to highlight the rear workbench and diffusors could provide sufficient working illumination, as well as a dramatic touch. Similar track lighting could provide working illumination at the workstation.

Background noise level

This room is intended primarily for postproduction work, hence recording of production sound might be rare. The primary need is that noise be kept at an appropriately low level so that the operator can evaluate sound from the monitors with no interference.

Video projection

A possible video screen position is indicated in the north elevation
of Fig. 7-7. A projector position near the operator's desk is probably indicated. Keystone screen distortion can be minimized by tip-
ping the screen top toward the projector. The projector, video
monitors, and other equipment on the desk are all possible cul-
prits in producing early sound reflections of a type to distort the
sound the operator hears. As a last resort, absorbing blankets over
the top and back of desk equipment might be necessary to control
these reflections.

8

Home theater

NEVER HAVE THE PLANETS OF SOPHISTICATED EQUIPMENT and program software been so aligned for the home theater as they are today. The stores are awash with audio amplifiers with very effective processors for surround sound, video projectors that challenge cathode-ray type television receivers, and video libraries of a scope and quality unimagined in the past. High-definition television promises a major upgrade of the television image quality, which has been static since the 1950s. Sound and picture quality available on a routine basis to the consumer compares favorably with the best in theaters.

Things have been happening in the acoustical area as well. For one thing, the average person is beginning to realize that the link between the loudspeaker diaphragm and his/her ear is every bit as important as the quality of the microphone, amplifier, loudspeaker, and talent. This acoustic path is mysterious to many, less tangible and less available for correction and adjustment than all the audio/video hardware and software combined. Acoustic experts are looked upon as true gurus. The home theater needs and should have excellent audio, but that is impossible without an excellent acoustical treatment of the space between the source and the ear.

The home theater design to be presented here includes a heavy dose of sound diffusion. The discovery in the 1960s of the application of number theory to practical sound diffusing devices to hang on the wall is an amazing story. Today *quadratic-residue*, *primitive-root*, and other diffusors with even fancier eyebrow-raising names have increased the quality of sound recording in practically every major recording studio in the world. For the designs of this book, the author is deeply indebted to those diffusor pioneers as the wealth of experience they have accumulated is the key to a broad and expansive sound, a new sharpness and clarity of the stereo sound field, as well as a sense of being thoroughly immersed in the music and sound effects.

Space for the home theater

If the interest is in truly simulating the values of the motion-picture theater, more space is required than for a high-fidelity listening situation. As a point of departure a dedicated space for the home theater will be assumed and then, at the end of the chapter, possible compromises will be considered to trim the theater idea down to a more affordable size.

Whether the modest theater space considered here will be new construction or an adaptation of an existing space, there is need to consider room proportions. First-class acoustics is a goal as well as first-class visuals, and the distribution of room resonance frequencies is the logical first step in good acoustics. Small rooms are by far more difficult acoustically than larger ones, because of the dominance of modal resonance frequencies that are too few and far between. Special attention is required to control them.

Theater plan

The floor plan of Fig. 8-1 shows a room of 8'-0" × 12'-10" × 18'-7", which follows the ratio of 1.00 × 1.60 × 2.33 (Table 12-3).

A study of the modal frequencies of this room below 300 Hz is given in Table 8-1. Glancing down the difference column reveals a rather coarse but well-distributed field of axial modes. The average spacing is almost 15 Hz, and there are numerous spacings in the 20-Hz region. It is evident that the axial modes are spaced as well as can be expected for a room this size. Low-frequency absorption and much diffusion should make the room an acceptable listening room.

Figure 8-1 shows the placement of vital furniture and pieces of equipment. The two main loudspeakers are arranged in a normal hi-fi style, except that the sofa is placed somewhat behind the "sweet spot" to widen the sweet spot a bit. Adding diffusion to the room will further widen the area of good perception of the stereo image, extending it over the entire sofa.

The 10-foot distance to the screen requires a picture larger than the usual picture tube for optimum viewing, hence a video projection system is suggested. Placing the video projector in front of the viewers is rejected, because sound reflections from the projector would distort the sound. At the rear of the room sound reflections from the video projector will actually contribute to the room am-

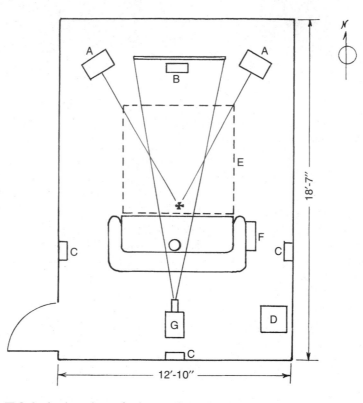

■ 8-1 *A plan view of a home theater.*

bience. A five-channel surround processor is needed, plus (A) a left stereo loudspeaker, (A) a right stereo loudspeaker, (B) a center loudspeaker, (C) surround loudspeakers, and (D) a subwoofer and (F) a possible control position. The positions of the loudspeakers are critical except for the subwoofer (D). At the low bass frequencies that it radiates, directional effects are minimal.

Early reflections and their effects

To avoid confusion, another floor plan (Fig. 8-2) is introduced for acoustical details. Also for clarity, some of the items in the floor plan of Fig. 8-1 are left out.

To simplify, four sound rays from each stereo loudspeaker will be considered. Sound ray #1 from the left loudspeaker is reflected from the left side wall to the listener's ears. Sound ray #2 from the left loudspeaker travels to the rear of the room and will be discussed later. Sound ray #3 from the left loudspeaker hits the right side wall and is reflected directly to the listener. Sound ray #4 from

■ Table 8-1 Axial-mode distribution (room: 8'0" × 12'8" × 18'6")

	L L = 565/18.6	W W = 565/12.8	H H = 565/8.0	Ascending order	Diff.
f_1	30.4 Hz	44.1 Hz	70.6 Hz	30.4	13.7 Hz
f_2	60.7	88.3	141.3	44.1	16.6
f_3	91.1	132.4	211.9	60.7	9.9
f_4	121.5	176.6	282.5	70.6	17.7
f_5	151.9	220.7	353.1	88.3	2.8
f_6	182.3	264.8		91.1	30.4
f_7	212.6	309.0		121.5	10.9
f_8	243.0			132.4	8.9
f_9	273.4			141.3	10.6
f_{10}	303.8			151.9	24.7
				176.6	5.7
				182.3	29.6
				211.9	0.7
				212.6	8.1
				220.7	22.3
				243.0	21.8
				264.8	8.6
				273.4	9.1
				282.5	21.3
				303.8	
				Mean	14.7 Hz
				Std. dev.	8.8

the left loudspeaker, generated at the corners of the cabinet by diffraction, travels from the corner of the cabinet to the front wall and is then reflected to the listener. Passing over ray #2 which goes to the rear of the room, concentrate on rays #1, #3, and #4, all of which reach the ears of the listener indirectly.

These early reflections reach the listener, but at slightly different times because of the different paths followed. All three rays carry the same program material, they just arrive at different times. These rays combine with each other and with the direct ray. When replica signals combine with time differences between them, "comb filters" are formed; these change a flat response to one having alternating peaks and valleys down through the spectrum.

When two signals add in phase, a peak is produced; when they add in phase opposition, they cancel, creating a null. The combining of these early sound reflections at the ear of the listener adds comb filter distortion to the direct signal. To get rid of the distortion, the

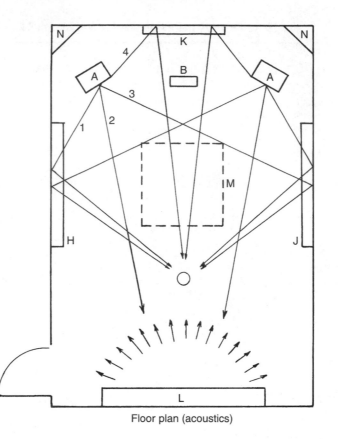

Floor plan (acoustics)

■ **8-2** *Plan view home theater showing reflections.*

early reflections must be eliminated or greatly reduced (see Figs. 6-1 and 6-2).

Controlling early reflections

Our goal, then, is to produce a reflection-free zone where the listeners sit, so they can hear the sound coming from the loudspeaker without the distortion of the early reflections. This is done by mounting panels of absorber on the side walls at strategic points where they will intercept rays #1 and #3 (Fig. 8-2). Ray #4 will require another such panel (K) on the front wall behind the loudspeakers.

Without indulging in too much detail, reflections from both loudspeakers from the floor require an absorber on the floor (E, Fig. 8.1), and those from the ceiling necessitate another absorber (M) on the ceiling. If properly placed and of the proper material, these absorbing panels should reduce the amplitudes of the early reflections and protect the listener(s) from their distortion.

The absorbing panels for the side and front walls (H, K, J) are specified as RPG Abffusors, which are not only excellent absorbers (0.80 at 100 Hz), they are also good quadratic-residue diffusors. These are mounted flat on the wall with no extra air space.

The ceiling absorber specified (M) is a $4' \times 4' \times 2''$ RPG Ceiling Cloud (M), which is suspended $4''$ below the ceiling. To reduce the amplitude of the floor reflections, what could be more practical than an attractive $6' \times 6'$ rug (E, Fig. 8.1)?

Ray #2 (Fig. 8-2) and its many counterparts from both loudspeakers travel to the rear of the room where they encounter a panel (L) of 8 RPG 734 Diffusors (Fig. 12-15). The four upper diffusors diffuse sound vertically, and the four lower ones diffuse sound in the horizontal direction.

The many rays #2 sweep past the listener and are reflected from room surfaces on their way to the rear of the room. Some of this energy strikes the diffusors, some the wall. It is returned to the listener either in specular or diffuse form.

Other theater details

Figure 8-3, showing the east and west elevations, illustrates clearly the floor and ceiling early reflections and the absorbers for controlling them. RPG B.A.S.S. Trap low-frequency absorbers (N) are mounted behind the screen, one in each of the four tricorners.

This bass absorber (Fig. 5-11) has a peak absorption that coincides with the first-order 70.6-Hz axial mode associated with the height of the room. It is less effective for the modes associated with the length (30.4 Hz) and width (44.1 Hz) of the room, but there is still good absorption at these frequencies. The location of the bass absorbers in the corners at one end of the room places them where all modes terminate, thus in the most effective position. Should the room be judged "too boomy," more bass absorbers should be mounted between the existing ones.

The side-wall diffusors (J and H) and the diffusor behind the screen (K) are made of two $2' \times 2' \times 4''$ RPG Abffusors below for horizontal diffusion, and one $4' \times 2' \times 4''$ Abffusor above for the vertical diffusion. The diffusor behind the screen is identical to the two side-wall diffusors.

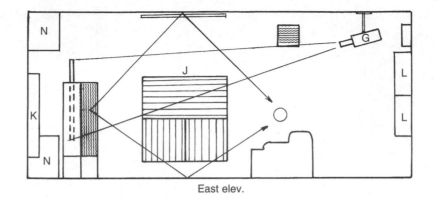

East elev.

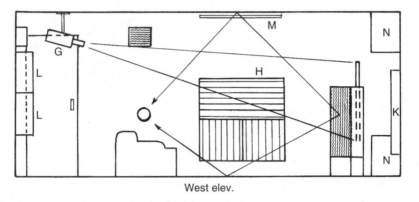

West elev.

■ 8-3 *East and west views of a home theater.*

The north and south elevation sketches are shown in Fig. 8-4. The south elevation shows four RPG Abffusors (L) in the top row positioned for vertical diffusion. The four on the bottom row are positioned for horizontal diffusion (L). In the south elevation sketch, a possible location for the subwoofer (D) is indicated on the floor in the southeast corner of the room.

Both main stereo loudspeakers require bases to bring them up to a desirable height. These should be simple 3/4″ plywood boxes, painted black and filled with sand to deaden the cavity resonance.

There will be numerous pieces of electronic equipment associated with this home theater. None of this equipment should be placed in front of the listener, where it would produce its own private early reflections. A short rack located behind the sofa is probably the wisest location. In that position it can be reached over the back

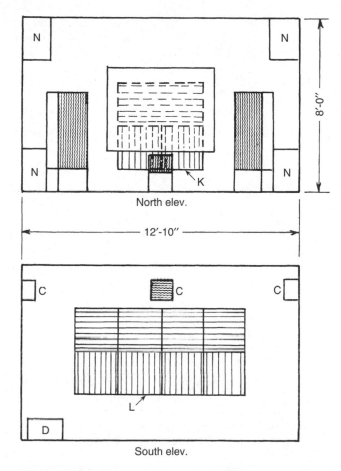

North elev.

South elev.

■ **8-4** *North and south views of a home theater.*

of the sofa or a control panel could be located at the right end of the sofa as shown in Fig. 8-1.

The listening environment

Now that these early sound reflections have been cared for, what can be expected in the quality of the listening situation? Once the early reflections are eliminated (or at least well attenuated), the stereo field will be sharp and definite. The location of the instruments in a musical group becomes clear. The sound stage becomes filled with three-dimensional images and the listener(s) feel immersed in the music.

The center loudspeaker duplicates the situation in the commercial theater. It helps enlarge the stereo sweet spot for central images, and the rest of the sound field is improved as well. The center loudspeaker makes the stereo field more clear and speech more intelligible.

The contribution of the subwoofer (D) is that the very-low-frequency components are boosted. Perhaps the greatest effect is in building up the drums in certain types of music. Sound effects are the greatest beneficiary, from the clicks of a closing car door to explosions.

Reverberation time

To this point the acoustic effort has been in the direction of reducing the effects of early reflections. Will the room be too dead? Experience has shown that for a room of this size (1900 ft^3), a reverberation time of 0.3 to 0.5 seconds would be reasonable. Table 8-2 lists the details of reverberation time calculation, and Fig. 8-5 gives the results.

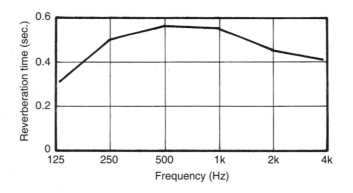

■ **8-5** *A reverberation time graph for a home theater.*

Without even planning for it, the reverberation time is within reason. The gypsum board (drywall) walls overabsorb the low frequencies. A concrete floor is assumed, or a wood floor may add a bit more to low-frequency absorption. The conclusion is that there is no point in straining toward a certain reverberation time at this stage.

A better approach would be to measure reverberation time when construction is completed and the room furnished. With this information it can be decided whether the room is bright enough for music and dead enough for speech intelligibility.

■ Table 8-2 Reverberation calculations (Home theater)

Description	Area sq ft, S	125 Hz		250 Hz		500 Hz		1 kHz		2 kHz		4 kHz	
		a	Sa	a	Sa	a	Sa	a	Sa	a	Sa	a	Sa
Drywall	740	0.29	214.6	0.10	74.0	0.05	37.0	0.04	29.6	0.07	51.8	0.09	66.6
Abffusor	80	0.82	65.6	0.90	72.0	1.00	80.0	1.00	80.0	1.00	80.0	1.00	80.0
B.A.S.S.	16	0.40	6.4	0.20	3.2	0.10	1.6	0.05	0.8	0.03	0.5	0.02	0.3
Rug 6 × 6	36	0.02	0.7	0.06	2.2	0.14	5.0	0.37	13.3	0.60	21.6	0.65	23.4
Ceiling cloud (assume 703 mtg 7)	16	0.66	10.6	0.95	15.2	1.00	16.0	1.00	16.0	1.00	16.0	1.00	16.0
Sofa (est.)	60	0.30	18.0	0.35	21.0	0.40	24.0	0.50	30.0	0.60	36.0	0.65	39.0
Total Sa			315.9		187.6		163.6		169.7		205.9		225.3
Reverberation time (sec.)			(0.30)		(0.50)		(0.57)		(0.55)		(0.45)		(0.41)

$$\text{Reverb. time} = \frac{(0.049)(1305)}{\text{Total } Sa}$$

The compromise home theater

Not everyone interested in a home theater can afford a special space for it. Here are a few suggestions for compromise and the penalty accompanying each compromise.

☐ The video projector could be replaced by an ordinary television receiver and its smaller images.

☐ If a woofer is available, the demands on the main stereo loudspeakers are reduced, making possible the use of smaller, less-expensive loudspeakers.

☐ Eliminating the five-channel sound processor and the surround feature entails a large step down, but home theater of a sort is possible without these enhancements.

☐ It is possible to forget all these early reflections and just live with the distortion, as most high-fidelity owners do.

The world has lived for years with only the diffusion offered by surface irregularities, convex surfaces, etc., without number-theory reflection-phase-grating diffusors, but who wants to go back? The home theater is a statement for the future. Compromises are statements for the past.

The home listening room

9

THE VARIETY OF HOME LISTENING ROOMS IS SO GREAT THAT little generic discussion toward their improvement can be possible, except to provide a basic understanding of the acoustical principles involved in such rooms. Further encouragement to take this tack is the impression that acoustics of the listening room is far less understood than the operation of amplifiers and loudspeakers. For these reasons, room modes and the effect of reflections on the quality of sound will be the main course, with a few examples of specially designed listening rooms as the dessert.

The lower frequencies

Where to start in improving the home listening room? It seems logical to start at the bottom (the bottom end of the audio spectrum, that is). The low-frequency end of the spectrum is the source of a large portion of the problems normally encountered. At the mid- to high- or high-frequency end of the spectrum the wavelength is relatively small, and it is acceptable to think of sound as rays bouncing around the room. At 300 Hz, the wavelength of sound is 1130 ft/sec divided by 300 cycles/sec, or 3.8 ft per cycle. At 100 Hz the wavelength is 1130/100 = 11.3 ft.

Your subwoofer probably delivers significant energy at 50 Hz, at which the wavelength is 1130/50 = 23 ft. So, at 100 Hz the wavelength of sound is about equal to the width of your room, and at 50 Hz it is equal to the length. The ray concept below about 300 Hz has no meaning. In the lows, the room dimensions are of the same order as the sound being delivered by the loudspeakers. The room has now changed from a room full of sound rays (above about 300 Hz) to a space that resonates at many audible frequencies (below 300 Hz). Because the transition between the two regions is a gradual one, the use of 300 Hz above is only a convenience. Chapter 18 could be helpful here.

Modes in typical rooms

The following rooms will be examined closely:

Small apartment:	$19'-6'' \times 10'-6'' \times 8'-0''$
Medium-sized room:	$24'-0'' \times 14'-6'' \times 9'-0''$
Large-sized room:	$27'-9'' \times 17'-0'' \times 9'-6''$

Of course, there are larger and smaller rooms (cut up in various ways) that are used for listening environments, but there is much to learn from these three. In Table 9-1 the frequencies of the axial modes below 300 Hz have been calculated for the small apartment. All of these are active resonances ready to carry low-frequency music components from the loudspeakers to the listeners.

■ Table 9-1 Axial-mode study
(Small apartment: $19'-6'' \times 10'-6'' \times 8'-0''$)

	L $L = 565/19.5$	W $W = 565/10.5$	H $H = 565/8.0$	Ascending order	Diff.
f_1	29.0 Hz	53.8 Hz	70.6 Hz	29.0 Hz	24.8 Hz
f_2	57.5	107.6	141.3	53.8	3.7
f_3	86.9	161.4	211.9	57.5	13.1
f_4	115.9	215.2	282.5	70.6	16.3
f_5	144.9	269.0	353.1	86.9	20.7
f_6	173.8	322.9		107.6	8.3
f_7	202.8			115.9	25.4
f_8	231.8			141.3	3.6
f_9	260.8			144.9	16.5
f_{10}	289.7			161.4	12.4
f_{11}	318.7			173.8	29.0
				202.8	9.1
				211.9	3.3
				215.2	16.6
				231.8	29.0
				260.8	8.2
				269.0	13.5
				282.5	7.2
				289.7	29.2
				318.9	
				Mean	15.26
				Std. dev.	8.9

The lowest frequency to enjoy modal support in this room is 29 Hz, which is the first-order mode associated with the length of the room. In other words, sound energy of 29 Hz will receive a resonant boost as it is reflected back and forth between the front-end and back-end walls. The sound pressure over the surface of the end walls will be high, and there will be a vertical null plane at the center of the room (it would be instructive to inspect Fig. 5-2 at this point). If this null plane coincides with a listener's head at the sweet spot, he would perceive no 29-Hz component of the signal. Walking to the rear or front walls, however, 29-Hz energy would be intense.

Table 9-1 shows not only the first-order frequency of 29.0 Hz, but also 53.8 Hz (the first-order mode between the two side walls) and 70.6 Hz (the first-order mode of the resonance between the floor and the ceiling). The integral multiples of these first-order frequencies (second-, third-, fourth-orders, etc.) are just as hot as the first-order frequencies. The array of three columns of frequencies of Table 9-1 constitutes the "acoustics" of the room below 300 Hz.

Mode spacing

Everyone loves a smooth response curve. The acoustic response curve of the room of Table 9-1 below 300 Hz can be imagined by plotting the entire 19 narrow response curves on a linear frequency scale. All 19 have approximately the same amplitude. It would be fair to assume a bandwidth of 4 Hz for each resonance, measured between the −3 dB points. A glance at such a plot would emphasize the need for uniform *distribution* of resonances, which determines the *spacing* between adjacent resonances.

Evenness of distribution of resonances along the frequency scale is determined by the proportions of the room. In a cubic room the length, width, and height first-order resonances would be identical, each multiple resonance would have triple strength, and the spacing between modal resonances would be great. Careful proportioning of length, width, and height serves to distribute the resonances more uniformly, as described in Table 12-3.

In Table 9-1 all the modal frequencies below 300 Hz are arranged in ascending order, and from these the spacing of adjacent modes is determined by calculating the difference in frequency between adjacent modes. The spread of differences is from 3.3 Hz to 29.2 Hz, but the average difference is 15 Hz. Sixty-seven percent of the differences are within 8.9 Hz of 15 Hz. The latter is called the standard deviation, which is of great interest to statisticians but of limited value in this context. Studies of audible colorations of speech have shown that 15-Hz separation of adjacent modes is usually acceptable, but spacings greater than 25 Hz and zero spacing (called a coincidence) can both cause colorations.

Effects of room size

Tables 9-2 and 9-3 for a medium-sized and a large-sized room give the opportunity of studying the effect of size on mode spacing on sound quality.

Table	Size	Mean Mode Spacing	Std. Dev.
9-1	Small	15.26 Hz	8.9 Hz
9-2	Medium	12.29	7.05
9-3	Large	9.25	8.67

■ **Table 9-2 Axial-mode study**
(Medium-sized room: 24'-0" × 14'-6" × 9'-0")

	L L = 565/24.0	W W = 565/14.5	H H = 565/9.0	Ascending order	Diff.
f_1	23.5 Hz	39.0 Hz	62.8 Hz	23.5 Hz	15.5 Hz
f_2	47.1	77.9	125.6	39.0	8.1
f_3	70.6	116.9	188.3	47.1	15.7
f_4	94.2	155.9	251.1	62.8	8.0
f_5	117.7	194.8	313.9	70.6	7.3
f_6	141.3	233.8		77.9	16.3
f_7	164.8	272.8		94.2	22.7
f_8	188.3	311.7		116.9	0.8
f_9	211.9			117.7	7.9
f_{10}	235.4			125.6	15.7
f_{11}	259.0			141.3	14.6
f_{12}	282.5			155.9	8.9
f_{13}	306.0			164.8	23.5
				188.3	0.0
				188.3	6.5
				194.8	17.1
				211.9	21.9
				233.8	1.6
				235.4	15.7
				251.1	7.9
				259.0	13.8
				272.8	9.7
				282.5	23.5
				306.0	
				Mean	12.29
				Std. dev.	7.05

	L L = 565/27.67	W W = 565/17.0	H H = 565/9.5	Ascending order	Diff.
f_1	20.4 Hz	33.2 Hz	59.5 Hz	20.4 Hz	12.8 Hz
f_2	40.8	66.5	118.9	33.2	7.6
f_3	61.3	99.7	178.4	40.8	18.7
f_4	81.7	132.9	237.9	59.5	1.8
f_5	102.1	166.2	297.4	61.3	5.2
f_6	122.5	199.4		66.5	15.2
f_7	142.9	232.6		81.7	18.0
f_8	163.4	265.9		99.7	2.4
f_9	183.8	299.1		102.1	16.8
f_{10}	204.2			118.9	3.6
f_{11}	224.6			122.5	10.4
f_{12}	245.0			132.9	10.0
f_{13}	265.4			142.9	20.5
f_{14}	285.9			163.4	2.8
f_{15}	306.3			166.2	12.2
				178.4	5.4
				183.8	15.6
				199.4	4.8
				204.2	20.4
				224.6	8.0
				232.6	4.9
				237.5	7.5
				245.0	20.4
				265.4	0.5
				265.9	20.0
				285.9	11.5
				297.4	1.7
				299.1	
				Mean	9.25
				Std. dev.	8.67

The tight 9-Hz average spacing in the large room underlines the acoustical value of large venues. The constancy of the standard deviation would suggest that similar factors of spacing are at work in all three sizes of rooms.

Low-frequency peaks/nulls in the listening room

Recall the array of low-frequency resonance frequencies in Table 9-2, as applied to the medium-sized listening room. It is not enough to imagine the three columns of frequencies spotted over the listening room floor. Each of those frequencies affects the entire listening room. Figure 9-1 is a feeble attempt to apply only the first three resonances of only the length and width columns to the room.

Taken from Table 9-2, the sound pressure lines for 23.5 and 47.1 Hz are shown at the top of Fig. 9-1, and 70.6 Hz at the bottom to avoid crowding. A sound-pressure peak is always present at the ends of the room and between nulls. Referring to the width column of Table 9-2, the frequencies of 39.0, 77.9, and 116.9 are selected and sketched onto Fig. 9-1. The nulls for these six resonances are indicated as broken lines on the floor plan. These lines are really the bottom edge of null planes reaching to the ceiling (see Fig. 5-2).

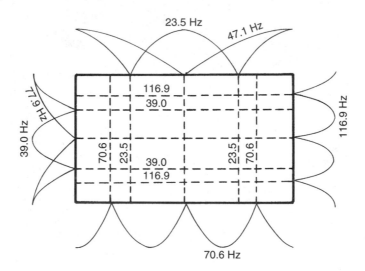

■ **9-1** *Some of the null-plane edges for medium-sized listening room.*

Only six of the 23 room resonances below 300 Hz from Table 9-2 are included in Fig. 9-1. If all 23 were plotted in Fig. 9-1, null-plane edges would cover the floor, and walls and peaks would be just as plentiful. There is no avoiding them, nor do we want to avoid them, but it would be nice if the nulls were not so deep and the peaks were not so high. That is what low-frequency absorption can accomplish.

Imagine sitting in the room of Fig. 9-1 with all peaks and nulls in place while a musical selection is being played. Different modes are excited as the musical notes go up and down the scale and chords are played, resulting in an acoustically scintillating room-space. If only our eyes could see it, how beautiful it would be! What an addition it would be to what the ear hears!

These modal peaks and nulls affect the listening at the sweet spot and the positioning of the loudspeakers. Proper low-frequency acoustical treatment of the room will tend to smooth out the peak-null differences, and will minimize the variation of response with position in the room. A room treated only with tile, fibrous, or foam material of 1-inch thickness will not affect the modes because these offer practically no absorption in the low-frequency region. Low-frequency absorption requires special techniques, devices, and effort.

Background noise

In an existing house, most of the noise producers are already in place. The central heating system is probably of the small-duct, high-velocity air type. The water pipes are fastened solidly to the structure so that their noise is efficiently radiated into the rooms. The noise of daily living is cheerfully generated by the occupants. In other words, there is very little that can be done except listen to the music at 2:00 A.M., and that is frowned upon. The lucky person might possibly be able to choose a room at some distance from the kitchen clatter, or have such an irresistible system that all the human noise makers are gathered to listen.

Positioning loudspeakers

There are a few tricks in loudspeaker location that might be in conflict with long-held ideas of propriety. For example, placing a loudspeaker symmetrically in a tricorner puts a 9 dB low-frequency peak in its response. Change this location from a tricorner to a spot on the floor close to a wall, and the low-frequency peak is 6 dB. Place it close to an isolated wall and the peak is only 3 dB. The loudspeaker interacts with reflecting surfaces near it. If your system needs the peak, try for a symmetrical arrangement that will give it. The peak cannot be defeated if there are walls anywhere around, but it can be minimized by placing the loudspeaker at different distances from each of the three reflecting surfaces.

Dipole loudspeakers, such as the electrostatic type (which have a tall, vibrating membrane), have a strong rear radiation component that must be controlled to prevent another source of early reflections. The more common magnetic-type loudspeakers radiate rearwards chiefly by the diffraction from corners of the cabinet.

Identification of early reflections

Figure 9-2 identifies all the reflection possibilities in the end of a rectangular structure. There are two situations in which these early reflections may become significant:

1. A person speaking into a microphone: The direct component from the speaker's mouth arrives at the microphone first, because it is closest. Soon after, reflected energy arrives from many surfaces. These early reflections interact with the direct and with each other, causing comb-filter distortion and degrading the quality of the combined signal.

2. A loudspeaker radiating sound to the ears of a nearby listener: The loudspeaker signal is reflected from many surfaces, and the reflections interact with the direct signal and with each other, creating comb-filter distortion and degrading the sound the listener hears.

The first-order (first-bounce) reflections are classified as early reflections because they are the first to arrive after the direct sound. The program quality perceived by the listener in a typical hi-fi situation is greatly affected by these early reflections. Assuming reasonable care in controlling the effect of the low-frequency modal resonances (the "boom"), and assuming play-back, amplifiers, and loudspeakers of reasonable quality, the early reflections are undoubtedly the major determinant of program quality at the "sweet spot."

Figure 9-2 shows how the effect of early reflections can be reduced by the use of absorbing material. The spot on the floor where sound from the tweeters of the two loudspeakers is reflected must be located. Laying a mirror on the floor and having it moved by an assistant until the listener can see the tweeter will locate a spot for each loudspeaker. A single rug (1) can easily cover both spots, and the floor bounce is controlled.

The position at which absorbent should be placed to control the left and right side-wall reflection (2) can also be found by using a mirror and an assistant. The location for the panels to absorb the ceiling reflections (3) is much more difficult to find with a mirror, but the experience on (1) and (2) probably suggest a good pair of spots, and success can be ensured by an absorbent of generous size. The absorbent behind the loudspeakers (4) can be logically located by knowing that the source is the edge of the loudspeaker cabinets.

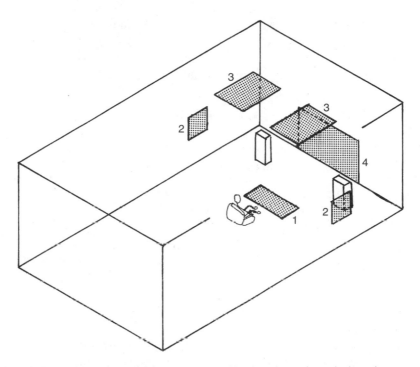

■ **9-2** *An isometric drawing showing the location of spot absorbers.*

Psychoacoustical effects of reflections

Even before controlling the early sound reflections by the above methods, it is urged that the condensed results of the research of Floyd Toole and Sean Olive (Olive 1989) on lateral reflections be reviewed in the section "Sound Reflections and Psychoacoustics" in chapter 18. These reflections effect the spaciousness impression, in which the listener feels immersed in the music. Shifting and spreading of the auditory image are also a function of reflection amplitude. In other words, one does not necessarily want to eliminate lateral reflections, but only to control them or to use them to adjust the degree of spaciousness perceived.

Acoustical treatment for the listening room

Caring for the early reflections is a big step toward the treatment of the listening room, but there are other things to do. It is very important to control the low-frequency modal resonances of the room. Absorbers must be placed in the corners of the room. Once the effect of placing low-frequency absorbers in two corners is heard, the desirability of doing the same in the other two corners will undoubtedly be evident. ASC Tube Traps may be used (see chapter 5 for their characteristics) or RPG B.A.S.S. Trap absorbers (see also chapter 5, especially Fig. 5-11).

The room should neither be dead nor too "bright." The normal living room reverberation time of about 0.5 second should be close to what is needed. If there is too much overstuffed furniture, it might be on the dead side. If the room seems too bright, fabric-covered glass-fiber panels could be introduced to strike the best balance.

Diffusion has been emphasized in every design of this book because it is such an important element. Diffusors on the rear wall of the listening room are encouraged. In fact, as we shall see later in this chapter, diffusors can control early reflections even better than absorption, because absorption does absorb more signal energy.

Examples of listening room treatment

This section on treating a listening room will have many similarities to chapter 4 on project studios. The reason for this is that a basic factor in working in a project studio is constantly checking sound quality and content. In chapter 4 the Abflector, the B.A.S.S. Trap, and the Skyline, products of RPG Diffusor Systems, Inc., were introduced as highly effective low-cost units that are especially attractive to the budget-limited person wanting high performance. The three listening-room designs to follow are really a single room with progressive additions to improve the performance of that room. Each design is acoustically optimized for that budget level.

Listening room A

The listening room of Fig. 9-3 treats the troublesome side-wall reflections on a minimum basis with a single Abffusor (1) on each side.

The ceiling reflection is controlled with two Nimbus ceiling clouds (2) which are thin fabric-covered absorbers spaced out from the ceiling to improve absorption. There is no treatment of the wall behind the loudspeakers at this budget level. Some control of low-frequency modes is offered by two B.A.S.S. Traps (3) on the rear wall. It is interesting that the peak of absorption of these low-frequency absorbers (70 Hz) coincides with the frequency of the first ceiling mode (575/8 = 70.6 Hz) for eight-foot ceilings, which are quite common in dwellings around the world.

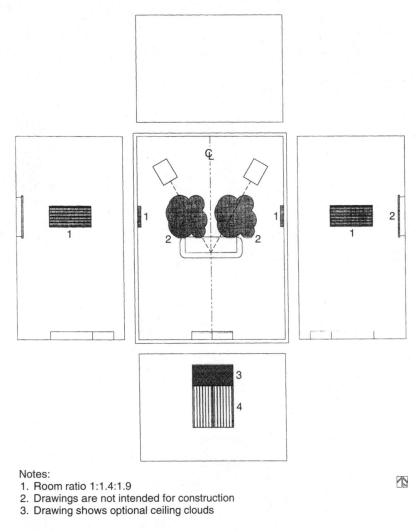

Notes:
1. Room ratio 1:1.4:1.9
2. Drawings are not intended for construction
3. Drawing shows optional ceiling clouds

■ **9-3** *Listening room A: Minimum treatment.*

Two RPG 734 QRD diffusing units (4) are also mounted on the rear wall. These are cost-effective wooden units based on prime 7 quadratic-residue theory, with a depth of 9″. This room should perform well, but not perfectly; there are other improvements reserved for rooms B and C.

Listening room B

When money is available to upgrade listening room A, the place to invest it is shown in Fig. 9-4. The only changes are a doubling of the area of the units on the side wall and the rear wall. Two Abffusors are now located on each side wall, and two more are added to the rear wall for a total of four. Two more B.A.S.S. Trap units are also added to the rear wall. Two more RPG 734 QRDs (4) have been added to the rear wall.

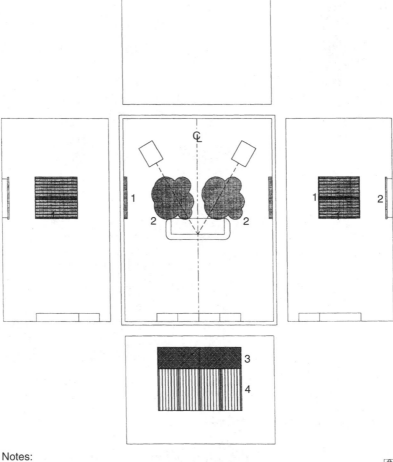

Notes:
1. Room ratio 1:1.4:1.9
2. Drawings are not intended for construction
3. Drawing shows optional ceiling clouds

■ **9-4** *Listening room B: Better treatment.*

Listening room C

The only change in going to level C (Fig. 9-5) is treating the wall behind the loudspeakers and adding low-frequency absorption. The front wall now has two Abffusors (5) as the outside units and two QRD 734s on the inside (4). Four B.A.S.S. Trap units (3) are below the four diffusors. Another absorber (not shown) could be a rug in front of the sofa to control the floor bounce.

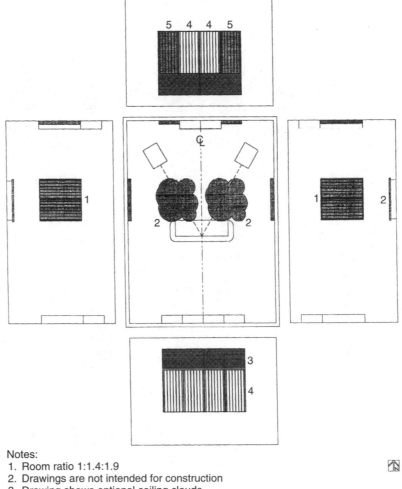

Notes:
1. Room ratio 1:1.4:1.9
2. Drawings are not intended for construction
3. Drawing shows optional ceiling clouds

■ **9-5** *Listening room C: Best treatment.*

With level C, the listening room is of excellent quality. The stereo image should be clear and definite over the entire sweet spot, and the sweet spot should be much larger. There should be a high degree of envelopment of the listeners in the music and, without the early reflections, sound quality should be excellent.

10

Teleconference room

FOR THE INDUSTRIAL OR ANY OTHER COMPLEX TO WORK, IT is important for people to get together to discuss and to plan. Having many people travel great distances is one way to do this, but the expense of travel and loss of time on the home job are so great as to encourage other methods. Today, audio and video facilities are readily available to bring people together irrespective of distance and at far less cost than traveling.

The availability of audio and video communication systems is one part of the solution; a space is also needed in which to use them. A dedicated space for in-house meetings and conferences is a useful and attractive idea. With only a slight extension of vision, such a local meeting room can serve as a teleconference room to bring in the rest of the world.

Requirements of a teleconference room

Speech intelligibility is by far the most important single need in a meeting room of any kind. Where speech arriving over telephone lines is to be heard, intelligibility is an even greater problem because of distortion in the transmission process. Room acoustics is the way to optimize speech intelligibility.

Speech is most intelligible in acoustically dead spaces. Intelligibility in highly reverberant spaces is very poor. Adequate acoustical design of a teleconference room involves many other things than sound absorption. The background noise level must also be low. It is well to keep the background noise level below the NCB-20 contour (see Fig. 12-8). To achieve this level of background noise, attention must, among other things, be given to HVAC (heating-ventilating-air conditioning) noise. A low velocity HVAC system is necessary, along with wrapped and lined ducts, and avoidance of certain fittings such as noisy air diffusers. Nearby noisy operations external to the teleconference room and the sound attenuation of the walls of the space must be brought into conformance. Once these are cared for, attention is directed to the treatment of the inside of the teleconference room.

Shape and size of room

The size of the room must be determined by the number of participants to be accommodated at one time. For 12 people plus the director, a 9'-0" × 14'-5" × 21'-0" space is selected; this fits the proportions 1.00 × 1.40 × 1.90. A listing of the normal modes of this space is given in Table 10-1. The difference column appears reasonable, with a single degeneracy at 188 Hz. The chance of this degeneracy causing an audible coloration are slim, because experience has shown that few colorations have been detected at this relatively high frequency, and the presence of tangential and oblique modes will tend to minimize the effect of the degeneracy. The volume of the room is 2722 ft^3.

■ Table 10-1 Axial-mode distribution (room: 21'-0" × 14'-5" × 9'-0")

	L L = 565/21.0	H H = 565/14.42	W W = 565/9.0	Ascending order	Diff.
f_1	26.9 Hz	39.2 Hz	62.8 Hz	26.9	12.3
f_2	53.8	78.4	125.6	39.2	14.6
f_3	80.7	117.5	188.3	53.8	9.0
f_4	107.6	156.7	256.1	62.8	15.6
f_5	134.5	195.9	313.9	78.4	2.3
f_6	161.4	235.1		80.7	26.9
f_7	188.3	274.3		107.6	9.9
f_8	215.2	313.5		117.5	8.1
f_9	242.1			125.6	8.9
f_{10}	269.0			134.5	22.2
f_{11}	296.0			156.7	4.7
				161.4	26.9
				188.3	0.0
				188.3	7.6
				195.9	19.3
				215.2	19.9
				235.1	7.0
				242.1	9.0
				251.1	17.9
				269.0	5.3
				274.3	21.7
				296.0	

Mean = 12.9
Std. dev. = 7.83

Floor plan

Figure 10-1(A) shows the basic plan of the suggested room and many of the acoustic elements. The importance of a special conference table cannot be overemphasized. A wedge shape is suggested so that each seated participant has a reasonably good view of the screen, or any activity at the head of the table.

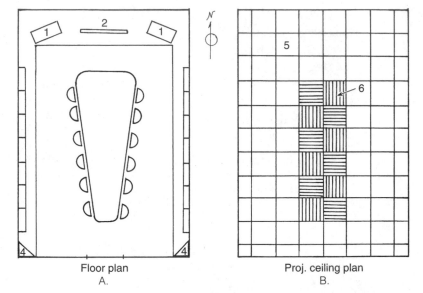

Floor plan
A.

Proj. ceiling plan
B.

Key: (1) Loudspeakers; (2) Video screen; (3) RPG Skyline diffusors; (4) RPG B.A.S.S. Traps; (5) Tectum Designer Plus ceiling panels; (6) RPG Abffusors; (7) Low-frequency Helmholtz absorber; (8) Carpet.

■ **10-1** *Floor/ceiling plans for a teleconference room.*

To reduce expense, the 30″-high shelf at the front of the room supports the loudspeakers. It also supports the screen and will probably be covered with transient equipment, such as overhead and slide projectors, tape recorders, and other equipment for multimedia use. Sliding doors make available for storage the space underneath this shelf. A 30″-high shelf continues along the sides of the room. To eliminate any possibility of flutter echoes between the east and west walls, and to provide supplemental diffusion for the room, 14 RPG Skyline diffusors are mounted on each side above the shelf. These are two-dimensional diffusors.

Ceiling plan

A standard T-bar suspended ceiling with a 16″ air space is shown in Fig. 10-1(B). In the center of this frame, directly over the table, are 14 RPG Abffusors (which both diffuse and absorb sound). The rest of the ceiling rack is filled with Tectum Designer Plus units having the dimensions 1-1/2″ × 24″ × 24″. Each Tectum panel is scored into 9 squares, which give a bold and interesting appearance.

Elevation views

The north, east, south, and west elevation sketches are shown in Fig. 10-2. These help to relate the two types of diffusors. All available wall space between the 30″-high shelf top and the suspended ceiling is covered with Tectum wall panel on a D-20 mounting (mounted on 3/4″ furring strips). The RPG Skyline diffusors are cemented to this Tectum wall panel.

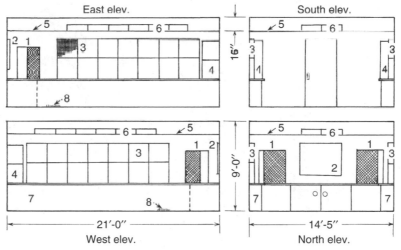

Key: See explanation of numbers (1) through (8) with Fig. 10-1 (p. 120).

■ **10-2** *Four elevation views of a teleconference room.*

In Fig. 10-2, south elevation, two RPG B.A.S.S. Traps (4) are placed on the 30″-high shelf on either side of the door. These are to help in controlling low-frequency effects caused by low-frequency modes. If such low-frequency effects persist, two similar stacks of the same bass units should be placed in the corners behind the loudspeakers. It is not expected that they will be needed.

A partial sectional view through a wall is shown in Fig. 10-3. The 30″-high shelf along the side walls is only about 12″ wide, to avoid intruding on the limited space of the room. On both sides of the room, beneath this shelf, a Helmholtz resonator low-frequency absorber is mounted. Its purpose will be described later.

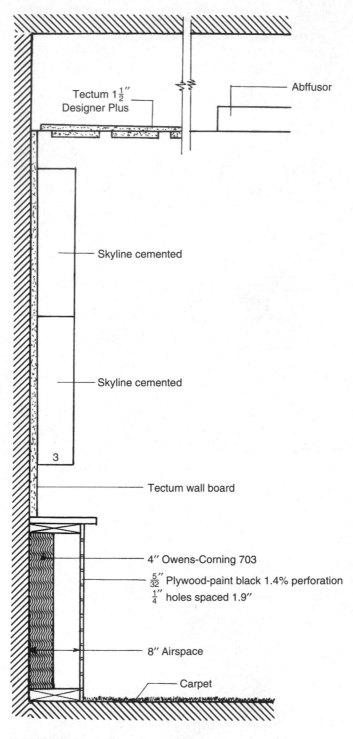

Tectum $1\frac{1}{2}''$ Designer Plus

Abffusor

Skyline cemented

Skyline cemented

3

Tectum wall board

4″ Owens-Corning 703

$\frac{5}{32}''$ Plywood-paint black 1.4% perforation $\frac{1}{4}''$ holes spaced 1.9″

8″ Airspace

Carpet

■ **10-3** *Wall section, teleconference room.*

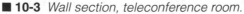

Teleconference room

■ Table 10-2 Reverberation calculations (Teleconference room)

Description	Area sq ft, S	125 Hz a	125 Hz Sa	250 Hz a	250 Hz Sa	500 Hz a	500 Hz Sa	1 kHz a	1 kHz Sa	2 kHz a	2 kHz Sa	4 kHz a	4 kHz Sa
Drywall, ½" 16" o.c.	940	0.29	272.6	0.10	94.0	0.05	47.0	0.04	37.6	0.07	65.8	0.09	84.6
Abffusor, ceiling	48	0.82	39.4	0.90	43.2	1.07	51.4	1.04	49.9	1.05	50.4	1.04	49.9
Tectum, ceiling	246	0.35	86.1	0.42	103.3	0.39	95.9	0.51	125.5	0.72	177.1	1.05	258.3
Designer Plus 1-1/2" × 24 × 24													
Carpet, heavy + pad	203	0.08	16.2	0.27	54.8	0.39	79.2	0.34	69.0	0.48	97.4	0.63	127.9
Tectum walls, D-20 mtg.	360	0.07	25.2	0.15	54.0	0.36	129.6	0.65	234.0	0.71	255.6	0.81	291.6
Total sabins			439.5		349.3		403.1		516.0		646.3		812.3
Reverberation time w/o comp. second.			(0.30)		(0.38)		(0.33)		(0.26)		(0.21)		(0.16)
Low-frequency compensation	97	0.80	77.6	0.90	87.3	0.38	66.0	0.28	27.2	0.18	17.5	0.12	11.6
Total sabins			517.1		436.6		469.1		543.2		663.8		823.9
Corrected reverberation time (sec.)			(0.26)		(0.31)		(0.28)		(0.25)		(0.20)		(0.16)

$$\text{Reverberation time} = \frac{(0.049)(2722)}{\text{Total } Sa}$$

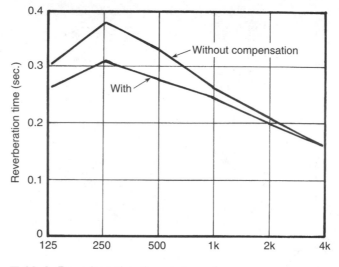

■ **10-4** *Reverberation time, teleconference room.*

Reverberation time

Reverberation time calculations are included in Table 10-2. The absorption units (sabins) are shown for six frequencies for the following elements: (a) the 12 ceiling RPG Abffusors, (b) the ceiling Tectum Designer Plus panels, (c) the Tectum wall panels, and (d) the heavy carpet with pad. These calculations lead directly to the plotted graph of Fig. 10-4 labeled "without compensation."

The reverberation time varies from 0.38 to 0.16 second, which is a reasonable range. A value of 0.2 second would represent a room with even greater absorption, which is just what is needed for good intelligibility. The room has adequate absorption at 125 Hz due to the diaphragmatic action of wall and ceiling gypsum board. The 0.38-second peak at 250 Hz suggests the possibility of adding a peak of absorption at this frequency. A Helmholtz unit having a peak about 250 Hz has been placed under the shelf. The 97 ft^2 available brings the reverberation time at 250 Hz down a bit, but not as much as desired. Even though it is impractical to equalize reverberation time to a reasonably flat 0.2 second, reverberation time between 0.3 and 0.2 second should make the room dead enough for good speech intelligibility.

Studio site selection

11

IT IS A TRUISM THAT THE QUIETER THE LOCATION OF THE proposed studio location, the easier it will be to achieve the required low noise level within it. The opposite of this is also a truism, and a very expensive one: The noisier the location, the heavier must be the walls, floors, and ceilings, and the greater the degrading effect of weak links such as windows, doors, and their seals. Frankly, a low background noise level within a recording studio or other audio space is so difficult and expensive to achieve structurally that it is imperative that much thought first be given to its location.

The ear is an adaptable organ. After working under noisy conditions for long periods of time, one gets used to the noise. Remember the lighthouse keeper who, when the foghorn suddenly stopped said, "What was that?" Microphones and amplifiers have no such adapting ability, and will truthfully register noise as noise. Thus a low noise level is a desirable goal for the studio, and one requiring considerable effort to achieve.

Any noise that gets through all the barriers is recorded on the track along with the desired signal. Some noises are more audible than others. The character of the noise is matched against the character of the signal, and the greater the difference between the two the more audible is the noise. For instance, consider a soft musical passage as the signal. Against that signal, the sound of a dog barking would be especially obnoxious, while that of a low-level random noise would be much less noticeable. Such reasoning is not of much comfort. Reducing all environmental and electronic noises to an absolute minimum must be our goal.

How low must the noise level be within the sound-sensitive space? It all depends upon how critical the work done in the studio will be. If noise standards are relaxed, there is no great problem. However, the demand for low noise levels is increasing all the time. For example, the extremely low noise levels common in digital recording and reproduction has focused attention on the noise at the pick-up point.

The sound level meter

The sound level meter, which is essential for the site noise survey, is actually a very simple instrument in principle. It incorporates a high-quality microphone to convert the sound pressure changes to voltage changes, a high-quality amplifier to amplify the voltage changes, a frequency-weighting network, a calibrated attenuator to extend the range, and an indicating meter. Obviously, the quality of the instrument depends on the quality of the integral parts and the stability of calibration.

The sound level meter accepts the noise disturbances in the air and reads out the sound pressure level. The sound pressure level is the sound pressure expressed in dB above the standard sound pressure of 20 micropascals, which is close to the threshold of hearing of the human ear. Remember, this is a physical measurement and is not a direct measure of loudness (which is a psychophysical parameter). However, loudness can be calculated from sound level measurements by involving several other concepts in the process. Sound pressure level is of great value in helping us manipulate the tangibles and intangibles of noise.

Most sound level meters incorporate three standard weighting networks, which change the frequency response as shown in Fig. 11-1. The purpose is to make the readings meaningful with respect to human perception of loudness. The A-weighting network

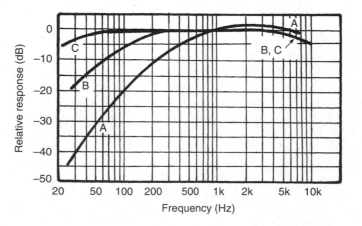

■ **11-1** *Weighting curves for sound level meters.*

is approximately the inverse of the 40-phon equal-loudness contour of the human ear. The B- and C-weighting networks are approximately the inverse of the 70- and 100-phon equal-loudness contours. It follows that the A-weighting network is intended for use in measuring low sound levels, B-weighting for intermediate sound levels, and C for high sound levels. These will make readings conform somewhat better with human judgments of loudness but, as mentioned before, sound level meters do not measure loudness.

Figure 11-2 shows two high-quality sound level meters, the Bruel & Kjaer Models 2236 and 2260. These are manufactured in Denmark. The Type 2236 is a precision integrating sound level meter. This instrument can display both peak and rms (root-mean-square) values simultaneously because it has two independent channels. The old-fashioned swinging-needle meter has been replaced by an LCD screen. An interactive dialogue guides the user through the measurements quickly and efficiently, and warns against changing the set-up parameters once a measurement has started. Three user-definable L-N values are available, and cumulative distributions of the results allow basic statistics on the spot. A real-time clock marks the date and time of every measurement. There is an automatic start feature that makes possible automatic starting of a run. L_{eq}, L_{max}, L_{10}, and L_{90} can be stored as a set, and 21,600 sets can be stored in the 128K memory. This is equivalent to six hours of logging at one-second intervals.

The Type 2260 is a precision sound analyzer based on the Type 2236 sound analysis software. It is extremely flexible because of the application software available. Up to a hundred meters of extension cable can be connected between the microphone/preamplifier assembly and the main body. The striking shape of the bodies of these two instruments has essentially eliminated distortions of the sound field due to physical interference of the instrument case. The instrument can be calibrated by a stable reference voltage available within the instrument, or an external sound level calibrator can be used.

Two other sophisticated sound level instruments are the Quest Model 1900 and the Larson-Davis Model 820, shown in Figs. 11-3 and 11-4. These manual or automatic, integrating, and data-logging sound level meters offer different weighting sound levels and 1/3 and 1/1 octave bands. It is possible to create and store multiple files without the necessity of downloading. The information is not lost if the instrument is turned off or the batteries removed.

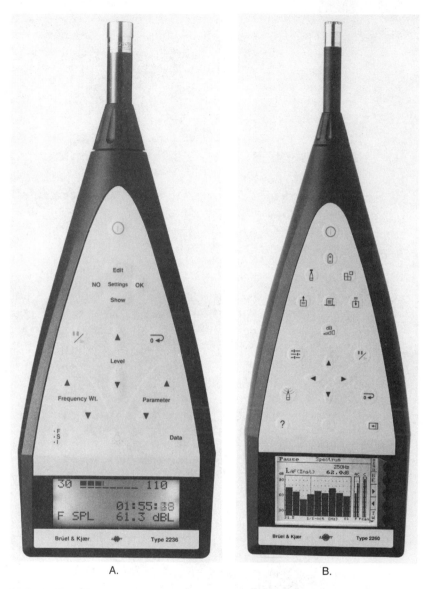

A.   B.

■ **11-2** *The Bruel and Kjaer sound level meters.*

These sound level meters are capable of handling almost any industrial, environmental, or community noise measurement problem.

Sound level meters of the sophistication of these four are not essential for a noise survey of a site, but they will make possible a very fast, complete, and accurate job. Measurements are so fast and simple that they will encourage taking measurements at more

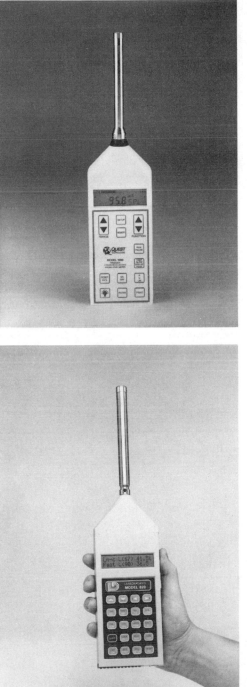

■ 11-3
The Quest sound level meter.

129

■ 11-4
The Larson-Davis sound level meter.

The sound level meter

environmental sites, or for extended hours or days for better coverage. A more complete definition of the noise exposure could only make the design of the sound-sensitive room that much better.

The noise survey

There are many economical aspects involved in choosing a site for the studio: land value, building cost, space rental, etc. It is also expensive to keep outside noises from getting into a studio. The smart move is to compare potential locations on the basis of 24-hour noise-level surveys to minimize construction costs. A noise-level survey is actually a rather simple operation, but one that yields extremely valuable information.

As a learning experience, there is nothing comparable to making an actual noise survey "by hand" with an old-fashioned sound level meter that provides only one thing: total sound levels. This might seem primitive, subelementary, and time-wasting, but it will give those running the survey an incomparable fundamental understanding of the process, and it is inexpensive.

First, a single point outdoors is selected that is close to the proposed studio site, and which seems to sample the known noisemakers of the neighborhood. A 24-hour stretch of measurements is then scheduled for this point. A sampling period could be hourly, or every 15 seconds. It is suggested that a measurement be taken every minute on the minute. This yields a set of 1440 samples, which should give a reasonable picture of the environmental noise of the neighborhood. Readings should be taken with the A-weighting network in the circuit. These are called dBA measurements.

A suggested data sheet system for recording measurements is shown in Fig. 11-5. Sound level readings will not be written down as numbers, but only as slashes in the appropriate column in which the levels fall. The fifth slash in any given column should be a diagonal to help in the counting later. Each column embraces 3-dBA readings, such as 48 to 50, 72 to 74, etc. Little final accuracy will be lost through this simplified 3-dBA coarseness. Every minute a slash will be added to this form in the appropriate column. This is the only duty until the 24-hour measurement period is completed.

At the end of the 24-hour period, there is sufficient data in hand to build a statistical distribution curve for that locality. The form on the upper part of Fig. 11-5 can now be filled in. The "Total" column is filled in, with the total number of slash marks in each column.

130

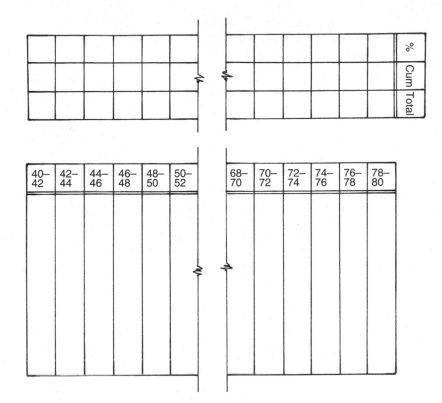

												%
												Cum Total

40–42	42–44	44–46	46–48	48–50	50–52		68–70	70–72	72–74	74–76	76–78	78–80

■ **11-5** *A form for collecting field data.*

When the "Total" column is full, entries can be made in the center "cumulative" column. The total number of slashes in the 78-80 column is added to the total of the 76-78 column, and recorded in the associated box in the "cumulative" column. The total of the 74-76 column is then added to the sum of the 76-78 and the 78-80 totals, and so on. The last "cumulative" entry will be the total of all slash marks on the page.

Now for the "percent" column. The cumulative figure in the first box is now divided by the total number of slashes, divided by 100 to express the figure as a percent, and the result is recorded in the first percent box. The second cumulative amount is then divided by the total in the cumulative column, expressed as percent, and recorded in the second box in the "percent" column, and so forth.

Everything is now in hand to draw a distribution curve such as Fig. 11-6. The first point on the distribution curve is found by plotting the first percent value in the "percent" column against the 79-dBA level. The second plotted point is the second percent value against

77 dBA, and so on. This is a significant curve that gives a complete statistical picture of the noise situation at that point, for that day in the week.

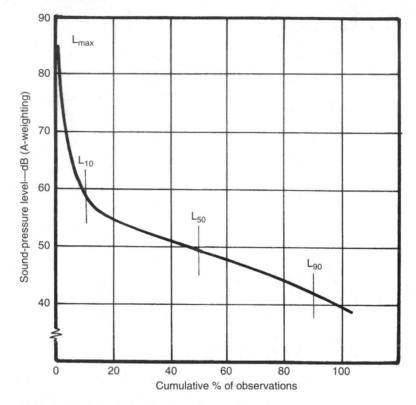

■ **11-6** *Statistical distribution of sound levels.*

In recording the data of Fig. 11-5, there will be times when airplanes fly overhead, trains pass, and sirens wail, giving levels far above the usual values. The maximum values must be recorded elsewhere, as the noise at this point is not complete without them. From all these extra high values, one is the highest and it becomes L_{max} for the survey. It is possible that this is the most important reading taken because it will penetrate the walls of the studio. All of the maximum values together can be analyzed as a separate problem. If that train passes several times during 24 hours, perhaps this is the one noise that dominates decisions of wall construction. On the other hand, if the plane passing overhead is the only one heard for a week, it should influence the wall construction less.

Figure 11-6 is indefinite on the L_{max}, but is very definite on L_{10} (the noise level exceeded 10% of the time), L_{50} (the noise level exceeded 50% of the time), and L_{90} (the noise level exceeded 90% of the time). This data is very important in deciding how strong the studio wall must be.

The question will arise: "How will L_{10}, L_{50}, and L_{90} vary throughout the day?" This is where the new integrating instruments excel. Figure 11-7 illustrates one way of presenting such information.

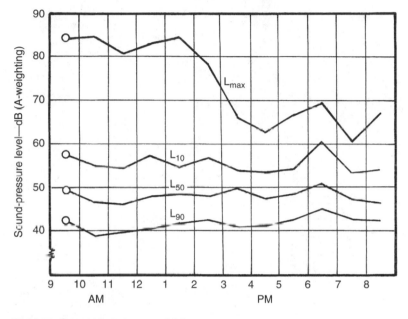

■ **11-7** *Sound levels over 24 hours.*

Noise criterion curves

The proper and the easiest way to define the noise level within any structure is to use the noise criterion curves of Fig. 11-8. These are the refined NCB curves (Beranek 1989) which, for many purposes, should be used instead of the older NC curves. For one thing, the NCB curves are extended downward two octaves below NC to the octave band centered on 16 Hz. Another difference is the inclusion of the shaded areas A and B. Sound-pressure levels of 70 or 80 dB magnitude below 63 Hz and which fall in areas A and B might induce audible rattles or feelable vibrations in lightweight partitions and ceiling structures. Other than these factors, the NCB and the NC curves are very close together.

During the noise survey, the sound level meter is used to measure the sound-pressure level in each octave band. The noise is described by this series of sound-pressure-level measurements in octave bands centered on 16 Hz to 8 kHz. These sound-pressure

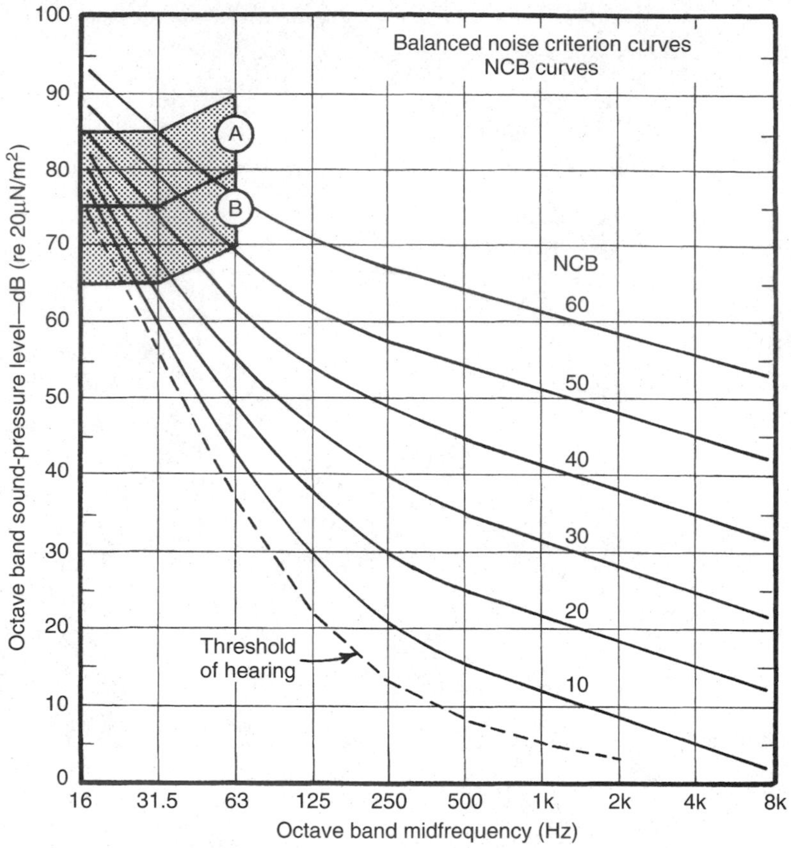
Balanced noise criterion curves
NCB curves
Octave band sound-pressure level—dB (re 20µN/m²)
100
90
80
70
60
50
40
30
20
10
0
A
B
NCB
60
50
40
30
20
10
Threshold of hearing
16
31.5
63
125
250
500
1k
2k
4k
8k
Octave band midfrequency (Hz)

■ **11-8** *NCB curves.*

levels are then plotted on the family of noise criterion curves, such as the black spots on Fig. 11-9. The sound can then be described by the highest criterion curve touched by the spots. The sound-pressure level in the 1 kHz octave band touches the NCB-20 curve, and can thus be described in simplified terms as an NCB-20 noise level. In this way, different noise levels can readily be compared on a single-number basis.

Relating noise measurements to construction

The highest noise levels are probably produced by airplanes flying overhead. Such intermittent, high-level noises tend to dominate the situation. To prevent such occasional noises from polluting an important recording, walls must be built strong enough to reduce such noises to a tolerable level. Figure 11-10 shows the spectrum of the jet noise as it passes over the studio. The lower curve is the

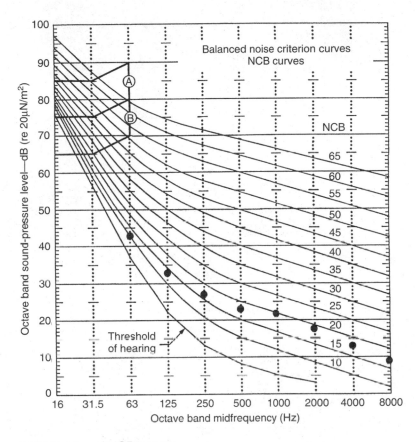

■ 11-9 *Use of NCB curves.*

NCB-15 contour, which is either the measured background noise of the studio or the studio noise goal, depending on the stage of the operation at which this study is made. The difference between the two curves, in dB, is the attenuation that the studio walls must provide.

Figure 11-11 is the difference between the jet noise curve and the studio noise curve of Fig. 11-10. This is the attenuation that the studio walls must deliver to bring the outside jet noise down to the desired background noise inside the studio.

What kind of walls are required to give the attenuation of Fig. 11-11? This is the subject of chapter 13, which must be anticipated slightly to understand the noise problem. The light, broken line in Fig. 11-11 is the Sound Transmission Class STC-60 contour. Walls rated as STC-60 in chapter 13 are of the more effective type. That class of walls offer approximately 60 dB of attenuation at 500 Hz. Without belaboring the subject, this is the general method of solv-

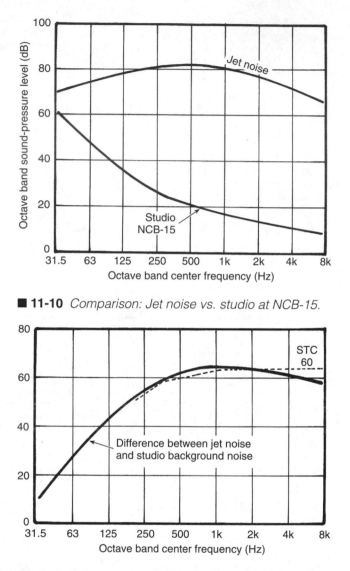

■ **11-10** *Comparison: Jet noise vs. studio at NCB-15.*

■ **11-11** *Difference between exterior and interior noise.*

ing the internal background noise problems of studios and other sound-sensitive spaces.

Locating a studio

Rarely are studios and audio rooms in isolated buildings with no noisy neighbor problems. Usually it is necessary to deal with other tenants, some of whom might be serious noisemakers in their own right. For example, is there an offset printing activity down the hall? A machine shop? A carpenter shop with a screaming buzz saw? An automobile repair facility? Such special circumstances require special solutions, and special solutions often require a consultant in acoustics.

Locating a space within a frame structure

A space for a studio or other sound-sensitive activity located within a frame structure will have the thud/thump problem of footfalls to contend with (elaborated in chapter 14). These low-frequency components are inevitable because of the limit to their control provided by lightweight frame structures. As long as the natural period of vibration of the floor-ceiling structure is about the same frequency as the peak energy of the thuds and thumps of footfalls, the footfall noise will be present and might actually be amplified. Carpeting stairs and hallways will reduce the higher-frequency components of footfalls, but the thud/thump low-frequency components will tend to penetrate floating floors, heavy carpet, or any other precautionary layer. If the type of work being done is not critical, extended low-frequency response might not be required. In such a case a high-pass filter can be inserted in the signal circuit, which would reject both signal components and troublesome noise below about 63 Hz (see Fig. 14-2). Extreme isolation measures to eliminate these thud/thump problems in frame structures are an exercise in futility.

Locating a space within a concrete structure

Locating a studio or other sound-sensitive area within a concrete structure does offer some prospect for effective isolation against the low-frequency components of footfall thuds and thumps. Although quite expensive, it is possible to build a studio-within-a-studio offering maximum isolation from outside and structurally-borne noise using familiar approaches (see chapter 3).

It is not wise to lay too much emphasis on the perfect ultra-quiet studio. After all, much of the world's audio and visual work is accomplished on a daily basis in environments that are less than perfect. To get the last few decibels of quietness can be a very expensive operation.

Insulation versus isolation

The distinction between *insulation* and *isolation* is basic, and should be clarified at this point. A given wall construction will offer a certain transmission loss to sound traveling through it. Sound transmission loss can properly be referred to as *insulation*. The studio under construction will have many sound transmission loss figures applying to it; that for the wall, the floor, the ceiling, the observation window, etc. The word *insulation* can be applied to the sound transmission loss of each element.

Isolation is more a global term applying to the studio as a whole. The difference between the sound-pressure levels taken inside and outside would be a measure of the true *isolation* of the studio. True isolation is what is wanted; numerous sound transmission losses (insulations) is the way isolation is achieved.

Diffusion of sound

12

DIFFUSION, IN THE CASUAL ACOUSTICAL SENSE, MEANS MIX-
ing of sound, making it more homogeneous. The physicist would
say more precisely that sound is diffuse when the energy density
of the sound is everywhere uniform, and all directions of energy
flow are equally probable. The more diffuse the sound in a recording studio or other audio room, the better (for most situations).
There are numerous ways of increasing the diffusion of sound. In
the past, reflection and absorption of sound have been the most
prominent variables in the acoustical design of an audio space. In
the studio designs of this book, however, diffusion takes on its
rightful priority. Leaving absorption of sound for later treatment,
let us consider reflection and diffusion, two separate entities that
are irrevocably related.

Diffusion by geometric shapes

Sound in a studio may be diffused by reflections off protuberances
of various shapes, such as those illustrated in Fig. 12-1. The flat
wall surface of Fig. 12-1(A) reflects sound in a specular manner,
with the angle of incidence equal to the angle of reflection. The triangular shape of Fig. 12-1(B) and the rectangular shapes of Fig.
12-1(C) reflect incident sound in various directions. This tends
toward mixing of the sound, but to a very limited degree. To be
effective, such irregularities must be large compared to the wavelength of the incident sound. Taking the speed of sound to be
1130 ft/sec, the wavelength of sound at 1000 Hz is 1.13 ft. A rectangular protuberance standing out from the wall two feet (about
two wavelengths) would be a poor diffuser below 1 kHz. Archi-
tects might find triangular, rectangular, or oddball shapes useful as
sound diffusers and visual features in large music halls, but space
limitations and limited effectiveness tend to rule them out for
most studio use.

Any sharp corner diffuses sound to a certain extent, in addition to
reflecting sound in various directions depending upon its shape. In

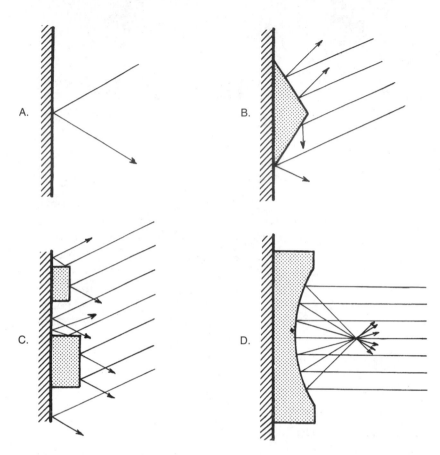

■ **12-1** *Patterns of reflection.*

this sense, there is a modest true diffusion from such edges as door openings [Fig. 12-1(A)], triangular shapes [Fig.12-1(B)], the rectangular protrusions of Fig. 12-1(C), or even the corners of Fig. 12-1(D).

The concave shape of Fig. 12-1(D) tends to concentrate the sound at a focal point. Concentrating sound is the opposite of diffusing sound; hence, concave shapes are very unpopular with those designing studios and other audio rooms with good (read diffuse) conditions.

The polycylindrical diffuser

The semicylindrical surface of Fig. 12-2 has something special to offer. Three things can happen to a plane wave of sound striking such a cylindrical surface:

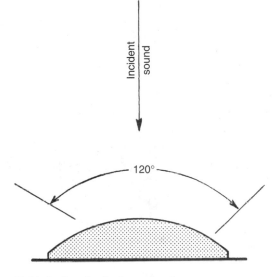

■ 12-2 *Semicylinder reflection.*

1. Some of the sound is absorbed.
2. Some of the sound is reflected.
3. Some of the sound is reradiated.

That portion of the sound that is absorbed is largely in the low-frequency region, because of the usual large dimensions. Absorption at the lower frequencies is ordinarily much needed in a room, and often difficult to achieve. The space between the curved diaphragm of a polycylindrical (poly) diffusor and the wall on which it is mounted is usually broken up into smaller spaces of varying volume. Each space is resonant at some low frequency at which absorption peaks. As the sections are of different volumes, these absorption peaks will be spread over a range of low frequencies that together provide valuable absorption in the normally difficult low-frequency region.

That portion of the incident sound that is reflected from the poly face is dispersed in various directions. This reflected component of dispersed sound must not be confused with the reradiated portion, although both disperse sound through a wide angle.

That reradiated portion of the sound results from the vibration of the curved diaphragm excited by the incident sound. This vibration of the poly diaphragm results in the reradiation back into the room of some of the sound energy. This reradiated sound does not follow the regular reflection law of equal angles of incidence and re-flection. Rather, it is radiated almost equally throughout an angle of about 120 degrees (Volkmann 1941). A similar flat diaphragm radi-ates sound through a much smaller angle. The excessive use of sound-absorbing material to correct acoustical deficiencies of a space results in the loss of signal energy. With polys, both the re-flected and the reradiated components conserve the precious sig-nal energy.

The first radio and recording studios used only drapes, carpet, and acoustical tile to control the reverberation time of the space. This resulted in overabsorption of high-frequency energy, and under-absorption of low-frequency energy.

The situation was favorably relieved through the increased popu-larity of polycylindrical diffusors, widely used in the 1940s and 1950s. At that time, some of the largest and best radio and record-ing studios incorporated a full complement of polycylindrical ele-ments on walls and ceilings. At this writing polys are not often seen in modern studios, although there is a definite trend toward their return. Their dramatic visual impact, low cost, and favorable acoustical characteristics are in their favor.

The room as a resonating chamber

There are other methods by which the diffusion of sound in a room can be enhanced; these are based on the concept of the room as a resonating chamber. Consider first a pair of opposing, parallel wall surfaces as illustrated in Fig. 12-3. These two isolated walls actu-ally are a resonating system. A tonal sound emitted between the two walls travels to the left wall and is reflected toward the right wall, then is reflected back toward the left wall, etc. The initial sound simultaneously travels also toward the right wall, is re-flected to the left, and then to the right. The right-going and the left-going waves combine, and if the wavelength of the sound is properly related to the distance L between the two surfaces, a standing wave will be produced as long as the sound continues.

The first mode of this standing wave has a maximum sound pres-sure at the surface of the walls, and a null midway between them. At twice that frequency, a second standing wave mode results with two nulls between the walls. At three times the frequency, three nulls appear. The sound pressure is maximum at the surfaces for all modes. This is called an axial mode, because the sound travels parallel to the axis normal to the two surfaces.

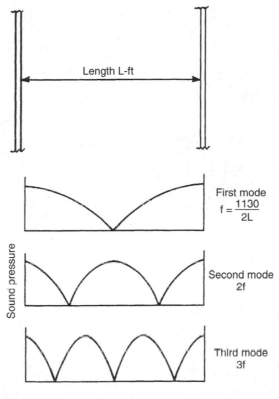

■ 12-3 *Two-wall resonance.*

Suppose the two opposing, parallel surfaces of Fig. 12-3 are the north and south walls of the enclosed space of Fig. 12-4. The axial mode path can be seen. However, there are other resonating systems in the room: one based on the tangential reflections, and the other on the oblique reflections. The tangential and oblique systems resonate in a way similar to the axial system.

The tangential mode involves reflections from four and the oblique mode all six surfaces. The added number of reflections reduce their energy. The axial modes, having the least number of reflections, are the more powerful. The tangential modes have only half the energy of the axial modes, which means they are down 3 dB (10 log 0.5). The oblique modes, having only one-fourth the energy of axial modes, are down 6 dB (10 log 0.25) (Morse and Bolt 1944).

For a rectangular space, such as the room of Fig. 12-4, there are three pairs of opposing, parallel surfaces, those associated with the length L, the width W, and the height H of the room. A solution

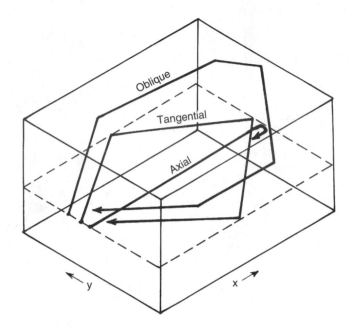

■ 12-4 *A comparison of axial, tangential, and oblique modes.*

of the wave equation for the frequencies of all the normal mode resonances is:

$$\text{Frequency} = \frac{c}{2}\sqrt{\frac{p^2}{L^2} + \frac{q^2}{W^2} + \frac{r^2}{H^2}} \qquad (12\text{-}1)$$

where

c = speed of sound, 1130 ft/sec
L,W,H = room length, width, height, ft
p,q,r = integers 0, 1, 2, 3, etc.

The integers p, q, and r are the only variables, as the length L, the width W, and the height H are fixed. The statement $p = 1$ refers to the first mode of Fig. 12-3, $p = 2$ refers to the second mode, $p = 3$ refers to the third mode, and so on.

Table 12-1 lists and identifies all the model frequencies for a small room, 12.46 × 11.42 × 7.90 ft, out to p, q, r = 4. Note that a frequency of 45.3 Hz is the lowest frequency supported by room resonance. Figure 12-5 might help you to visualize the pattern of sound pressure of the isolated 1,0,0 and the 3,0,0 modes of the room. Imagine the complexity if the sound-pressure pattern of all the modal frequencies of Table 12-1 were depicted simultaneously in Fig. 12-5! Combining only the 1,0,0 mode and the 3,0,0 mode is daunting, let alone all those listed in Table 12-1. The actual sound field is unbelievably complex, even in the simplest situation. As a loudspeaker emits the sound of an orchestra in a room, the sound pressure at any given point at any instant is the sum total of all these rapidly changing room modes.

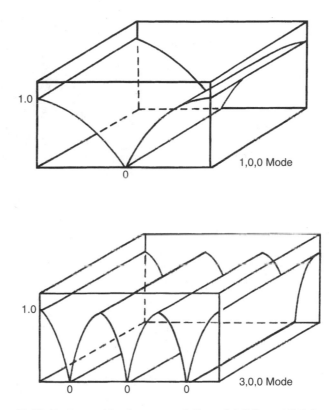

12-5 *A graphical representation of 1,0,0 and 3,0,0 modes.*

Another view of normal mode resonances in a room is given in Fig. 12-6. Here the axial, tangential, and oblique modes listed in Table 12-1 are each plotted against frequency. The spacing between these modes, especially adjacent axial modes, is a matter of importance. These axial modes tend to dominate the acoustics of the room. If one axial mode is separated too far from its adjacent neighbors it tends to act alone. Ideally, the coupling of adjacent modes is desirable for smooth room response.

The modal frequencies of Fig. 12-6 are depicted as narrow lines. Each mode has a bandwidth determined by the absorption in the

■ Table 12-1 Mode calculations
(room dimensions: 12.46 × 11.42 × 7.90 ft.)

Mode number	Integers p q r	Mode frequency, Hz	Axial	Tangential	Oblique
1	1 0 0	45.3	x		
2	0 1 0	49.5	x		
3	1 1 0	67.1		x	
4	0 0 1	71.5	x		
5	1 0 1	84.7		x	
6	0 1 1	87.0		x	
7	2 0 0	90.7	x		
8	2 0 1	90.7		x	
9	1 1 1	98.1			x
10	0 2 0	98.9	x		
11	2 1 0	103.3		x	
12	1 2 0	108.8		x	
13	0 2 1	122.1		x	
14	0 1 2	122.1		x	
15	2 1 1	125.6			x
16	1 2 1	130.2			x
17	2 2 0	134.2		x	
18	3 0 0	136.0	x		
19	0 0 2	143.0	x		
20	3 1 0	144.8		x	
21	0 3 0	148.4	x		
22	2 2 1	152.1			x
23	3 0 1	153.7		x	
24	1 1 2	158.0			x
25	3 1 1	161.5			x
26	0 3 1	164.8		x	
27	3 2 0	168.2		x	
28	2 0 2	169.4		x	
29	1 3 1	170.9			x
30	0 2 2	173.9		x	
31	2 3 0	173.9		x	
32	2 1 2	176.4			x
33	1 2 2	179.7			x
34	4 0 0	181.4	x		
35	3 2 1	182.8			x
36	2 3 1	188.1			x
37	2 2 2	196.2			x
38	0 4 0	197.9	x		
39	3 0 2	197.9		x	
40	3 3 0	201.3		x	
41	3 1 2	203.5			x
42	0 3 2	206.1		x	

Mode number	Integers $p\ q\ r$	Mode frequency, Hz	Axial	Tangential	Oblique
43	1 3 2	211.1			x
44	0 0 3	214.6	x		
45	1 0 3	219.3		x	
46	0 1 3	220.2		x	
47	3 2 2	220.8			x
48	1 1 3	224.8			x
49	2 3 2	225.2			x
50	2 0 3	232.9		x	
51	4 3 0	234.4		x	
52	0 2 3	236.3		x	
53	2 1 3	238.1		x	
54	3 4 0	240.2		x	
55	1 2 3	240.6			x
56	3 3 2	247.0			x
57	2 2 3	253.1			x
58	3 0 3	254.0		x	
59	0 3 3	260.9		x	
60	3 2 3	272.6			x
61	2 3 3	276.2			x
62	4 0 3	281.0		x	
63	0 0 4	286.1	x		
64	0 4 3	291.1		x	
65	3 0 4	316.8		x	
66	0 3 4	322.3		x	

room. The bandwidth of a mode is its width in Hz at points 3 dB down from the peak response. This width of a mode can be estimated by the statement 2.2/RT, where RT is the reverberation time of the space. Thus if the reverberation time of a studio is 0.5 second, the modal bandwidth is about 4.4 Hz. With a reverberation time of 1.5 seconds (such as an auditorium), the modal bandwidth would be about 1.5 Hz. These are sharp "tuning curves," but they are instrumental in bringing a degree of coupling of adjacent modes, due to overlapping of the skirts of the curves.

Early studies of colorations of sound in small studios of the British Broadcasting Corporation revealed that most voice colorations resulted from axial modes (or groups of axial modes) that were separated by more than 25 Hz. Such colorations were found in the approximate range of 75–200 Hz, with a subsidiary peak about 250–300 Hz. Speech colorations below 80 Hz are rare because of the low speech energy at those frequencies. These colorations dis-

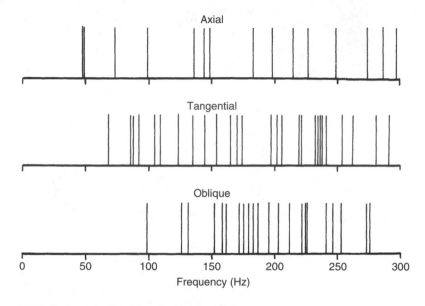

Axial

Tangential

Oblique

Frequency (Hz)

■ **12-6** *A plot of modes for Table 12-1.*

appear above about 300 Hz (Gilford 1972). Because of speech colorations, axial-mode separations greater than about 25 Hz are viewed with suspicion.

Figure 12-6 shows us that axial modes are not alone, even if they tend to dominate the situation. Tangential and oblique modes of lesser amplitude tend to "fill in" between the axial modes. In Table 12-2, the axial modes of Table 12-1 below 300 Hz have been separated out and arranged in a form convenient for analyzing the acoustical properties of this room. The dimensions of 12.46 × 11.42 × 7.90 ft tell us that this room is quite small. In the right-hand part of Table 12-2, the axial-mode resonances are arranged in ascending order and the differences between adjacent axial modes are listed. Several of these separations are greater than 25 Hz, which should serve as a warning of possible sound coloration in the 98–136 Hz and 148–181 Hz regions. The 24.7 Hz separation between 247 and 272 Hz would probably not cause colorations of speech because it is close to 300 Hz, the upper limit. The filling-in function of the tangential and oblique modes is also there to smooth the room response.

Diffusion by room proportions

If all three dimensions of a room are the same (the room is a cube), all three axial-mode frequencies would coincide, as well as their

■ Table 12-2 Axial-mode resonance frequencies of a "nonoptimum" room of 12.46 × 11.42 × 7.90 ft.

	Length L = 12′5½″ L = 12.46′ $f_1 = 565/L$	Width W = 11′-5″ W = 11.42′ $f_1 = 565/W$	Height H = 7′11″ H = 7.90′ $f_1 = 565/H$	Arranged in ascending order	Diff.
f_1	45.3	49.5	71.5	45.3	4.2
f_2	90.7	98.9	143.0	49.5	22.0
f_3	136.0	148.4	214.6	71.5	19.2
f_4	181.4	197.9	286.1	90.7	8.2
f_5	226.7	247.4	(357.6)	98.9	37.1
f_6	272.1	296.8		136.0	7.0
f_7	(317.4)	(346.3)		143.0	5.4
				148.4	33.0
				181.4	16.5
				197.9	16.7
				214.6	12.1
				226.7	20.7
				247.4	24.7
				272.1	14.0
				286.1	10.7
				296.8	

multiples. The same would be true of the tangential and the oblique modes. This would be a major departure from the goal of diffuseness. It is obvious that by making the three major dimensions of a room different, the cubical problem would be avoided and the modal frequencies would be distributed. The question is "How different?"

Choosing room dimension ratios for optimum distribution of modal frequencies was a favorite study in earlier days. A selection from the work of Ludwig Sepmeyer (Sepmeyer 1965) is given in Table 12-3. His mathematical analysis indicated that the room dimension ratios A, B, and C provide optimum (but not perfect) distribution of modal frequencies.

The small room analyzed in Table 12-1, Table 12-2, and Fig. 12-6 is quite nonoptimum, as the ratios of dimensions do not come close to those of Table 12-3. Studying the difference column in Table 12-4 shows that this room, following the B set of ratios of Table 12-3, has smaller differences than the nonoptimum room in Table 12-2. The basic procedure followed in the studio designs in the early part of this book is to use the ratios of Table 12-3.

■ Table 12-3 Rectangular room dimension ratios for favorable distribution of modal frequencies (Sepmeyer 1965)

	Height	Width	Length	
A	1.00	1.14	1.39	
B	1.00	1.28	1.54	
C	1.00	1.60	2.33	

Example: assume height of 10 ft.				**Volume**
A	10.00 ft	11.4 ft	13.9 ft	1743 cu ft
B	10.00	12.8	15.4	2035
C	10.00	16.0	23.3	3728

■ Table 12-4 Axial-mode frequencies of an optimum room
23.3 × 16.0 × 10.0 ft

	Length L = 23'-4" L = 23.3' $f_1 = 565/L$	Width W = 16'-0" W = 16.0' $f_2 = 565/W$	Height H = 10'-0" H = 10.0' $f_3 = 565/H$	Arranged in ascending order	Diff.
f_1	24.2	35.3	56.5	24.2	11.1
f_2	48.5	70.6	113.0	35.3	13.2
f_3	72.7	105.9	169.5	48.5	8.0
f_4	97.0	141.3	226.0	56.5	14.1
f_5	121.2	176.6	282.5	70.6	2.1
f_6	145.5	211.9	(339.0)	72.7	24.3
f_7	169.7	247.2		97.0	8.9
f_8	194.0	282.5		105.9	7.1
f_9	218.2	(317.8)		113.0	8.2
f_{10}	242.5			121.2	20.1
f_{11}	266.7			141.3	4.2
f_{12}	291.0			145.5	24.0
f_{13}	(315.2)			169.5	0.2
				169.7	6.9
				176.6	17.4
				194.0	17.9
				211.9	6.3
				218.2	7.8
				226.0	16.5
				242.5	4.7
				247.2	19.5
				266.7	15.8
				282.5	0
				282.5	8.5
				291.0	

Diffusion in nonrectangular rooms

There are those who would encourage the use of splayed walls and nonrectangular spaces to improve the diffusion of sound in a space. In the adaptation of an existing space, the splaying of walls is usually prohibitively expensive. In new construction, wall splaying can often be achieved at modest increase of cost.

There is no question as to the effectiveness of wall-splaying in the elimination of flutter echo. Flutter echo occurs when the space between two parallel reflecting surfaces is excited by some impulsive sound. The train of multiple reflections with its characteristic fluttering sound is very audible and disturbing. Splaying walls to eliminate flutter echo, however, is excessive. Flutter echo is easily eliminated by the strategic placement of a few patches of absorber or diffusor.

The hope of breaking up standing wave patterns in the space is another reason advanced for the use of nonrectangular rooms. In Fig. 12-7 the rectangular room is indicated by broken lines, and the skewed room by solid lines.

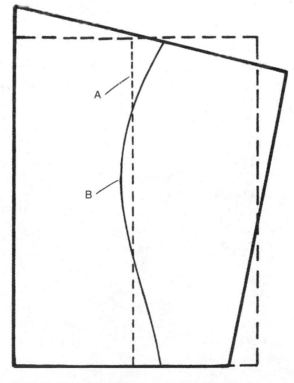

■ **12-7** *Skewing effects on mode distribution.*

In going from a rectangular room to its skewed counterpart, the null of a 1,0,0 mode is only distorted, not eliminated. Likewise, the normal-mode frequencies are all changed slightly, not eliminated. It is common to skew the heavy, outer walls of control rooms of recording studios. Whether this practice is justified is controversial. Some would prefer the predictability of the standing-wave sound field of a rectangular room over the intractable, distorted field of a skewed room.

Diffusion by reflection phase gratings

The past (the means of diffusing described above) is prelude to the future (the reflection phase grating diffusor). Diffraction gratings have long been the physicist's tool for studying light. Who has not been transfixed by the beautiful rainbow colors produced by a beam of sunlight falling on a prism or a diffraction grating? Edwin Hubble, at Mt. Wilson Observatory, used a diffraction grating made of a strip of glass on which a great number of closely spaced lines had been engraved. By allowing the light from distant stars to fall upon this grating, he noted that the frequency of red light from a star varied with the distance to the star. In this way, the concept of the expanding universe was born.

Acoustical gratings are now providing an ideal, highly effective means of diffusing sound. The grating diffusor has eclipsed all other means of diffusing sound in an enclosed space. To Manfred Schroeder (Schroeder 1975, 1979) goes the credit for the idea that groovy surfaces can make very effective sound diffusors (see Fig. 12-8).

The theory of the reflection phase grating sound diffusor is based on elementary (to the mathematician, that is) number theory to obtain the critical sequence of groove or well depths that diffuse incident sound through wide angles. Schroeder first experimented with maximum-length sequences, but quadratic residue, primitive (read prime) root, and other sequences have been shown to have superior properties.

Reflector theory

How do reflection phase grating diffusors really diffuse sound? First, consider the mechanism of a reflection from a plane surface. Why is sound reflected specularly, i.e., in essentially only one direction, by a flat surface? Consider that all points on the surface act like spherical emitters [see Fig. 12-9(A)].

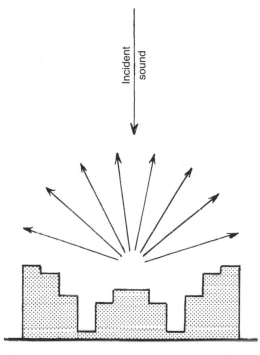

■ **12-8** *Diffusion of sound from a quadratic-residue diffusor.*

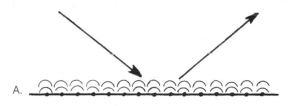

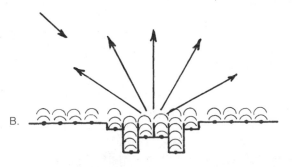

■ **12-9** *An explanation of how diffusors diffuse.*

The sound of all of these emitters is coherently combined to produce constructive interference (in which waves combine) and destructive interference (in which waves cancel). In the case of specular reflection, all scattered waves leaving the surface at an angle equal to the incident angle are relatively reinforced. This occurs because the specular waves differ in phase by integral numbers of wavelengths and hence they add. In other nonspecular directions, the waves cancel. Sound is actually scattered in all directions; the specular ones survive the interference process, and the nonspecular ones do not.

Reflection phase grating theory

How do reflection phase grating diffusors diffuse sound? Like the reflection case, the irradiated surface can be considered covered with point-source spherical emitters of reflected energy. Unlike the reflector case, we are interested in a way to alter phases of the scattered sound so that it can be constructively scattered in many directions, not just the specular direction. To do that, a prime 7 quadratic-residue reflector, as an example, can be sunk into the surface as in Fig. 12-9(B). The phases of the energy over the diffusor wells are different from that of the surrounding points because of the different depths of the wells. It takes time for sound to be reflected from the bottoms of the wells, and time is phase. With the different phases at the different wells, the energy that was directed only to the specular direction in the reflector case is now directed into many directions uniformly. The diffraction directions are determined by the width of the repeat unit, not the depth sequence.

Two periods of a typical diffusor are shown in Fig. 12-10 to clarify the definition of terms. Note the shared well between the two periods. Separators are advised for maximum efficiency, especially at oblique angles, as they preserve the acoustical integrity of the wells. The incident sound enters the "top" of the well. All of the wells are of the same width. The depth of the deepest well and the width of the wells determine the frequency range over which diffusion is effective.

Reflection phase grating diffusors are also absorbers. They provide useful low-frequency absorption below the lower limit of diffusion. Recently it has been proven theoretically that when wells of varied depth are involved, sound absorption occurs because of increased particle velocity flows from one well to another to equalize sound pressure on the face of the diffusor. This mechanism provides serendipitous low-frequency absorption.

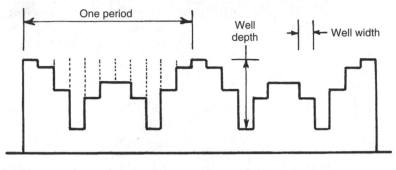

■ **12-10** *A graphical definition of terms used.*

It is good to pause for an overview of sound absorption and reflection as shown in Fig. 12-11. Much of the sound falling on absorbing material is absorbed, but a small amount is reflected. In the temporal response column the reflected sound is attenuated about 20 dB. In the spatial response column the angle of the reflection is equal to the angle of the incident ray, but the magnitude of the reflection is reduced by absorption.

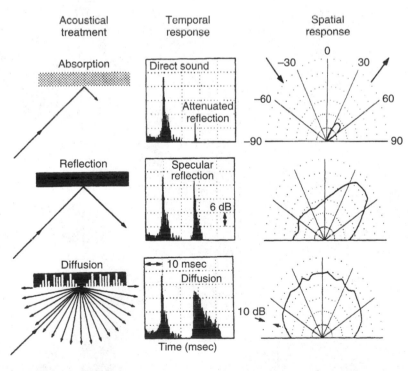

■ **12-11** *A comparison of sound incident on three surfaces.*

The reflection from a hard surface has almost the same amplitude as the incident sound. The reflection from the reflection phase grating diffusor, however, has several important points to emphasize:

1. The sound is scattered through the entire half circle, as shown by the spatial response.

2. The amplitude of the diffused return is attenuated approximately 8–10 dB.

3. The diffused energy is spread out over time.

These three points have very practical value, and will be referred to many times.

Figure 12-12(A) shows the hemidisc of sound diffused from the array of wells of a quadratic-residue diffusor. Figure 12-12(B) reminds us that sound is also reflected from the face of a phase grating diffusor in a specular manner, that is, as though it had a hard facing. One of the advantages of the primitive-root diffusor is that the specular scattering at the design frequency and multiples thereof are suppressed. This specular component exists and sometimes might cause problems, but in most installations it simply adds to the diffusion of sound in the room. Note particularly that the plane of specular reflection follows the "angle of incidence equals the angle of reflection" rule.

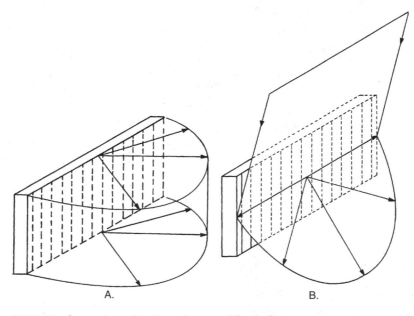

A. B.

■ **12-12** *Specular reflections from a diffusor face.*

The quadratic-residue diffusor

Of the several possible number theory approaches, the quadratic-residue sequence with its natural symmetry has proved to be of great practical value. The maximum well depth determines the longest wavelength of sound to be diffused. The well width is about a half wavelength at the shortest wavelength to be scattered. The relative depths of a sequence of wells are found from the equation:

$$\text{Well depth proportionality factor} = n^2 \text{ modulo } p \quad (12\text{-}2)$$

in which

p = a prime number
n – a whole number between zero and infinity

A prime number is any number that is not divisible without a remainder by any other integer (exceptions: +1, −1, and the integer itself). Examples of primes are 5, 7, 11, 13, 19, 23, etc. The *modulo* refers simply to residue. For example, inserting $p = 11$ and $n = 7$ into Equation 12-2 gives 49 modulo 11. Modulo 11 means that 11 is subtracted successively from 49 until the significant residue 5 is obtained. That is simple, but higher numbers such as $n = 15$ modulo 17 become awkward. The Hewlett-Packard HP-41C hand-held calculator has the modulo function in memory, making such computations simple. Also, in an appendix of Manfred Schroeder's book (Schroeder 1988), a modulo program is included. Figure 12-13 shows quadratic-residue profiles for p from 7, 11, 13, and 17.

The primitive-root diffusor

The number theory sequence for primitive-root diffusors is:

$$\text{Well depth proportionality factor} = g^n \text{ modulo } p \quad (12\text{-}3)$$

in which

p = a prime number
g = the least primitive root of p

Figure 12-14 shows primitive-root profiles for three combinations of p and g. Primitive-root diffusors lack symmetry which can be an advantage in some circumstances and a disadvantage in others. Their suppression of the specular scattering at the design frequency and multiples thereof is an important advantage of primitive-root diffusors.

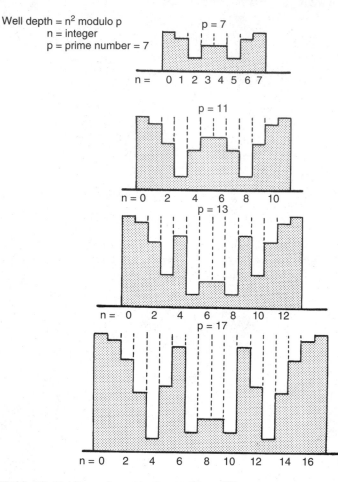

Quadratic residue

Well depth = n^2 modulo p
n = integer
p = prime number = 7

■ 12-13 *Profiles of quadratic-residue diffusors.*

Marketing reflection phase grating diffusors

Inventing something is one thing; getting the invention into people's hands in a successful and useful form is quite another. The credit for the original idea of the reflection phase grating acoustical diffusor, as previously mentioned, must go to Manfred Schroeder (Schroeder 1975). His insight into the mathematics upon which such diffusors are based and his knack for sensing practical uses are profound.

On the other hand, the successful commercialization of these ideas must go to Peter D'Antonio, a trained physicist with long scientific research experience. It is not unusual in the technical

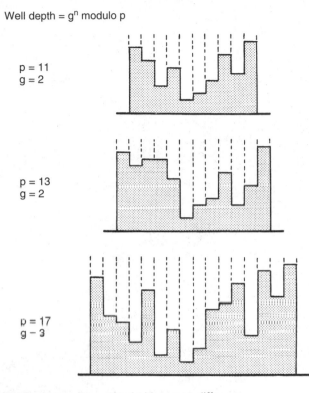

Primitive root

Well depth = g^n modulo p

p = 11
g = 2

p = 13
g = 2

p = 17
g = 3

■ 12-14 *Profiles of primitive-root diffusors.*

world we live in to find successful entrepreneurs who are trained scientists. Dr. D'Antonio approached this new idea in diffusion with energetic research, development, and prompt publication of the results. He founded RPG Diffusor Systems, Inc., of which he is president and chief executive officer, and started supplying practical diffusing elements to audio, recording, and architectural industries hungry for a device that, for the first time, offered good diffusion of sound. The studio designs of this book owe much to the availability of commercial reflection phase grating diffusors and to the generous cooperation of Dr. D'Antonio.

The QRD-734 quadratic-residue diffusor

One line of quadratic-residue diffusors offered by RPG Diffusor Systems, Inc., is shown in Fig. 12-15. These are designated the QRD-734 diffusors, based on the prime 7, and are available in nominal dimensions of 4′ × 4′, 2′ × 4′, 4′ × 2′, and 2′ × 2′ with a depth of 9″.

The width of the wells is 2.5″. They are available in birch veneer plywood, Melamine, or unfinished particle board. The well dividers are 0.50″ wood.

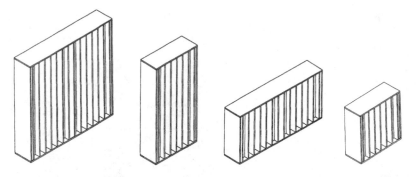

■ **12-15** *The QRD-734 diffusors.*

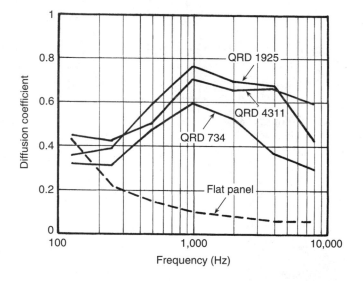

■ **12-16** *The diffusion coefficient of 734s.*

The performance of QRD-734 diffusors is best expressed in the graph of diffusion coefficient vs. frequency of Fig. 12-16. The QRD-734 also offers sound absorption coefficients varying in the range of 0.2 to 0.3. The only difference between the 734 QRD and the Diviewsor QRD is that the latter is made of plexiglass so that it is transparent.

The QRD-725/1925 quadratic-residue diffusor

Another line of diffusors offered by RPG Diffusor Systems, Inc., shown in Fig. 12-17, is the QRD-725 (based on prime 7) and QRD-1925 (based on prime 19), both having well widths of 2.5″. Perhaps the reader has already discovered the company's logical system of designating various types of diffusors: the QRD stands for quadratic-residue diffusor, of course, and the 725 means prime 7, well width 2.5″; 1925 means prime 19 and well width 2.5″.

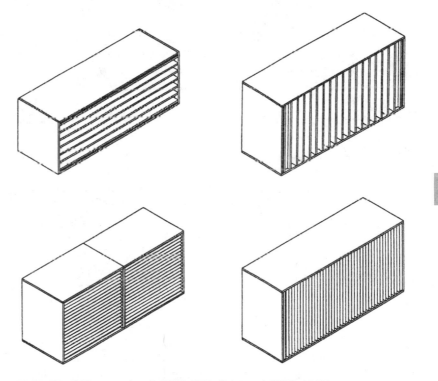

■ **12-17** *Diffusors: (top) QRD-725; (bottom) QRD-1925.*

The horizontal wells of the QRD-725 tell us that diffusion is vertical, the vertical wells of the QRD-1925 that its diffusion is horizontal. The diffusion coefficient of the QRD-1925 diffusor, Fig. 12-16, shows that it has excellent diffusion characteristics. Both the QRD-725 and the QRD-1925 have faces 4′ × 16″ with depths of 16″. Both have 0.50″ wood dividers. The options of birch veneer plywood, Melamine, and unpainted particle board apply also to the QRD-725 and the QRD-1925.

The diffusion coefficient

A very logical question concerning number theory diffusors is this: "Over what frequency range are they effective?" Peter D'Antonio and John Konnert (D'Antonio and Konnert 1984) have outlined the theory and application of reflection phase gratings. The low-frequency limit is imposed by the maximum depth of the wells, and the high-frequency limit is related to the width of the wells. The well width must be a half wavelength at the highest frequency to be effective as a diffusor, and the greatest well depth must be about 1-1/2 wavelength at the lowest frequency to provide adequate diffusion.

A maximum well depth of 12 inches and width of 1 inch yields a frequency range from 323 Hz to 5780 Hz. Experience has shown that the effective range is roughly about half an octave lower and half an octave higher than such figures. A far better appraisal of the true bandwidth is given by the graphs of diffusion coefficient of Fig. 12-16. Diffusion coefficient is the ratio of the scattered intensity at +/−45 degrees to the specular intensity. Diffusors can be compared to a flat panel in regard to diffusion effectiveness. This family of graphs show that it is generally easier to diffuse high-frequency sound than low-frequency sound, but the interesting thing is that all units shown have been designed to cover the audible band quite well.

The Formedffusor

The Formedffusor of Fig. 12-18 is a lightweight unit capable of providing broadband, uniform sound diffusion in either the horizontal or vertical planes for all angles of incidence by proper orientation of the units. Used in the standard T system as a suspended ceiling, the Formedffusor gives good mid- to low-frequency diaphragmatic sound absorption (as shown in Fig. 12-19). The diffusor is formed from Kydex by a thermoforming process, and uniform wall thickness is thus assured.

The FRG Omniffusor diffusor

The FRG Omniffusor, pictured in Fig. 12-20(A), provides broadband, uniform sound diffusion simultaneously in both horizontal and vertical planes for all angles of incidence. It is based on the two-dimensional quadratic-residue sequence with prime 7. The FRG Omniffusor is made of fiberglass-reinforced gypsum (FRG).

162

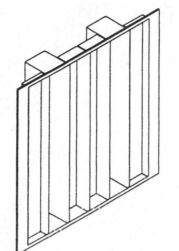

■ **12-18**
The Formedffusor.

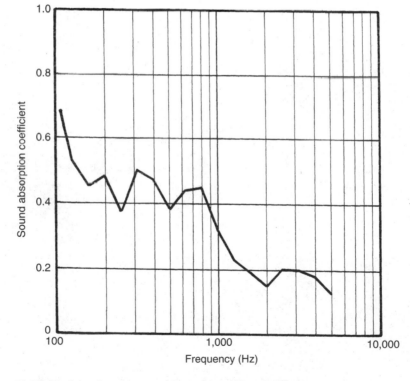

■ **12-19** *Low-frequency absorption of Formedffusor.*

This two-dimensional diffusor provides uniform hemispherical coverage from 500 to 3000 Hz. The nominal size of each unit is $2' \times 2' \times 5''$.

The Flutter-Free diffusor

Figure 12-20(B) shows the RPG Flutter-Free, which is a long, narrow strip of hardwood routed to form a 7 prime quadratic-residue diffusor. Available in 4-ft or 8-ft lengths, the width of each panel is 4″ and the thickness is 1″.

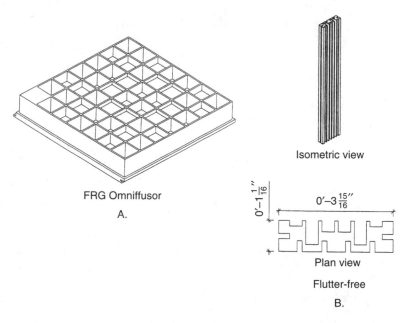

Isometric view

FRG Omniffusor

A.

$0'-1\frac{1}{16}''$

$0'-3\frac{15}{16}''$

Plan view

Flutter-free

B.

■ **12-20** *The Omniffusor and Flutter-Free diffusors.*

Our hearing is very sensitive to flutter echoes produced by successive reflections from opposing parallel surfaces. Such a series of impulses is perceived as a pitch or timbre coloration, which degrades sound quality and speech intelligibility. A typical waveform of a flutter echo is shown in Fig. 12-21. The use of patches of absorbent material has been the time-honored way of combating this echo. In some situations it is inadvisable to add absorption to a room to kill the flutter echo. Panels of Flutter-Free strips are very effective in killing flutter echoes by diffusion without adding absorption to the room.

Sound absorbers of the Helmholtz resonator type can utilize Flutter-Free slats. Such a low-frequency sound absorber is shown in

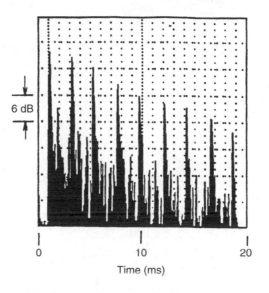

6 dB

0   10   20

Time (ms)

■ **12-21** *An echogram of a flutter echo.*

Fig. 12-22. The strips of Flutter-Free can be butt-joined with lamello splines, or joined together with a space of approximately 1.15″ with lamello spacers.

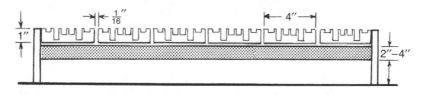

■ **12-22** *A Helmholtz absorber with Flutter-Free slats.*

Fractals

The self-similarity property of fractals has been applied to quadratic-residue diffusors by RPG Diffusor Systems, Inc. to extend their frequency range and minimize lobing for large-area coverage. There are certain manufacturing constraints that limit how narrow or how deep a diffusor well can be made. A similar problem with loudspeakers has been solved by woofers to extend the low-frequency radiation, and tweeters to extend the high-frequency radiation. Normal quadratic-residue diffusors can be designed for any reasonable low-frequency limit.

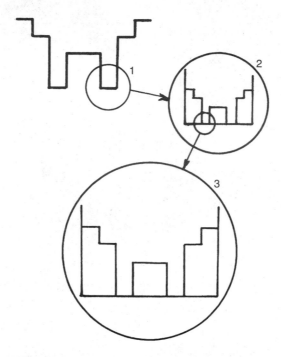

■ **12-23** *Applying the fractal principle.*

The fractal principle is used to extend the high-frequency limit. The general process is illustrated in Fig. 12-23. The upper left quadratic-residue unit is built on the prime 7. To extend its high-frequency range, another small prime-7 diffusor (circle #2) is fitted into the bottom of the indicated well in circle #1. In addition, smaller diffusors of the #2 type are fitted into every other well bottom of #1. Thus diffusor #1 acts in a normal way, except that the smaller #2 units in each well result in much greater high-frequency diffraction. That's not all, either; an even smaller prime-7 diffuser (#3) can be fitted into the bottom of each well of diffusor #2. So we have diffusors within diffusors within diffusors, each covering a different frequency range. A diffusor 16 ft long, 6 ft 8″ high, and 3 ft deep offers a bandwidth from 100 Hz to 17 kHz using the fractal principle.

13

Wall construction and performance

BEFORE WALL CONSTRUCTION CAN BE SPECIFIED, TWO MAJOR questions must be answered:

1. What noise level will define the environmental ambient noise in which the studio (or other sound-sensitive room under consideration) will operate?

2. What background noise level goal is to be set for the inside of the room?

The difference between these two noise levels defines the transmission loss the walls and other barriers must supply. This was the burden of chapter 11, and now this chapter will survey the construction and performance of walls as sound barriers.

The buzzer experiment

We can learn an important principle of sound isolation from a simple electric buzzer experiment suggested by Egan (Egan 1972). In Fig. 13-1(A) a sound level of 70 dB is measured by a sound level meter as the buzzer rests on the table. In (B), a box made of glass fiber is then placed over the buzzer and sealed. The sound level is still 70 dB inside the glass fiber box, and the outside level is reduced only slightly to 67 dB. In (C), a plywood box is next placed over the buzzer and sealed to the table. The inside sound level is now 75 dB, but the level outside has been reduced to 50 dB. In (D) the glass fiber box is slipped into the plywood box and the combination placed over the buzzer and sealed to the table. The sound level inside the box is now 71 dB, and the level outside is 43 dB.

The noise reductions in the four cases are tabulated in Table 13-1; they are (A): 0 dB, (B): 70 − 67 = 3 dB, (C): 75 − 50 = 25 dB, and (D): 71 − 43 = 28 dB. In (B), the glass fiber box offers only a tiny reduction of noise energy as it passes through the lightweight glass

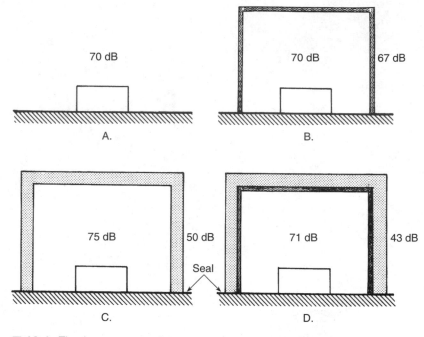

■ **13-1** *The buzzer experiment.*

fiber. In (C), the plywood box noise reduction of 25 dB is due to the much higher mass of the plywood as compared to that of the glass fiber in (B). A somewhat greater noise reduction results when the plywood box is lined with glass fiber in (D). The sound is trapped within the plywood box of (C) by multiple reflections, and the reverberant noise level within is increased from 70 to 75 dB. Lining the box with glass fiber reduces the reverberant noise level within

■ **Table 13-1 Buzzer experiment**

Condition	Sound level near bell	Sound level outside	Noise reduction
Bell in open	70 dB	70 dB	0 dB
Bell covered with box of ¾″ fiberglass	70	67	3
Bell covered with box of ½″ plywood	75	50	25
Bell covered with plywood box lined with ¾″ fiberglass	71	43	28

to 71 dB, giving a slight increase in noise reduction (to 28 dB, which is only 3 dB better than the plywood box alone).

Sound absorbents such as glass fiber absorb sound, but they are ineffective in blocking sound passing through them. The absorbents are low in density compared to the plywood and, for this reason, are quite ineffective as a sound barrier. Density (mass) is needed for high noise reduction. Any suggestion of reducing the noise from a neighbor's apartment by covering the shared wall with carpet or other absorbent should be rejected as futile.

Walls as effective noise barriers

A wall that is effective as a barrier to sound is characterized by a high transmission loss, i.e., the sound energy is decreased by passage through the wall. As transmission loss varies with frequency, complexity is introduced because all frequencies within the audible range are of interest to the audio person. To avoid the mistake of simplifying too soon, the graphs of Figs. 13-2, 13-3, 13-4, and 13-5 plunge the reader into the complexity of transmission loss curves of four common wall structures.

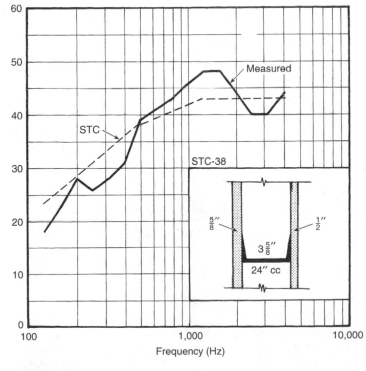

■ **13-2** *Transmission loss of a wall.*

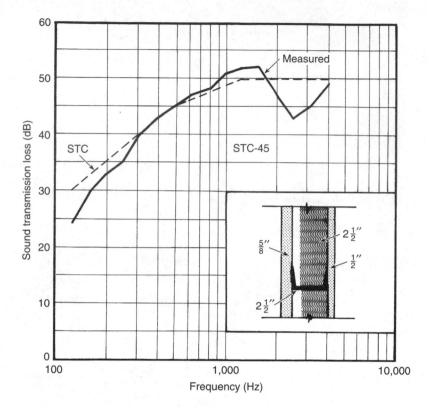

Sound transmission loss (dB) vs Frequency (Hz)

Measured

STC

STC-45

$2\frac{1}{2}''$

$\frac{5}{8}''$

$\frac{1}{2}''$

$2\frac{1}{2}''$

■ **13-3** *Transmission loss of another wall.*

It is hoped this procedure will provide substantial and logical reasons for many apparently arbitrary moves later in this chapter. Figure 13-2 (Northwood 1970) shows the measured transmission loss curve for the wall/partition pictured. This wall with 3-5/8″ steel or 4″ wood studs is faced with gypsum board ("drywall" or "hardboard") of 1/2″ and 5/8″ thickness. The shape of the transmission-loss curve is anything but regular, and describing it is a problem to the designer and the construction person. This problem has been simplified by the concept of the Sound Transmission Class or STC rating, which will be treated in the next section.

Sound Transmission Class (STC)

The Sound Transmission Class standard was introduced to rate the sound-blocking ability of a material or structure by a single number. It is a practical compromise for the sake of convenience. The measurement of the transmission loss of a partition, graphically presented as a function of frequency, is the elegant and ac-

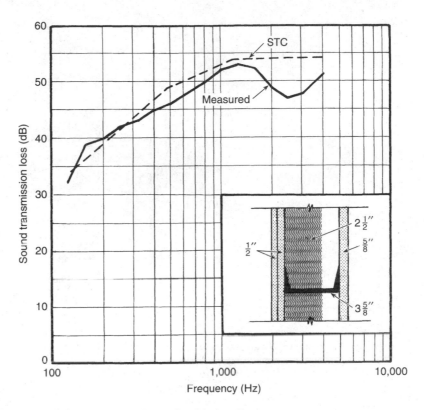

■ **13-4** *Transmission loss of a third wall.*

curate (and slow and costly) way to describe the partition's noise-blocking efficiency. Examples of such transmission loss graphs are shown in Figs. 13-2 to 13-5. The standard STC contour is shown in each figure as a broken line.

The shape of the standard STC contour is described by the data in Table 13-2. To use the STC contour, it is first plotted on translucent tracing paper to the same frequency and transmission-loss scales used in the transmission-loss graph. A vertical line at 500 Hz should be included. This tracing-paper overlay is then laid on the transmission loss curve, aligning the 500-Hz line of the overlay with that of the graph. The overlay is then shifted vertically with both 500-Hz lines together until the following conditions are met:

1. The maximum deviation of the test curve below the contour at any single test frequency shall not exceed 8 dB.

2. The sum of all the dB of the transmission loss curve variations at all 16 frequencies of the test curve below the contour shall not exceed 32 dB (an average deviation of 2 dB).

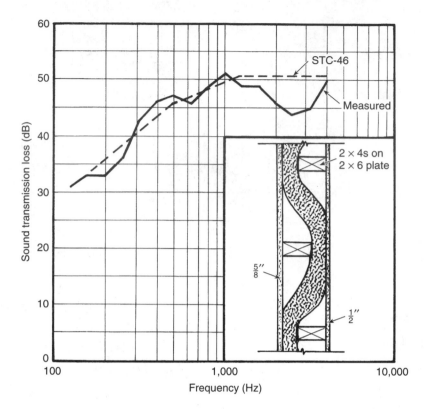

■ **13-5** *Transmission loss of a staggered-stud wall.*

When the position of the STC contour meets these two conditions, the STC value of the wall can be read off at the 500-Hz ordinate. The beauty of the STC contour is that it gives a convenient, single-number rating of the partition's ability to block sound that conforms reasonably well with practice and is widely accepted. With this shorthand method of evaluating walls and partitions, it is easy to compare one structure with another with regard to their ability to block sound. The STC-38 of the wall in Fig. 13-2 can then be compared with the STC-49 of Fig. 13-4, and there is some sense in saying one offers 11 dB more transmission loss than the other.

Not many of those associated with a studio construction project will need to handle STCs, except for the designer, yet the client needs to keep an eye on the designer!

Coincidence effects

An examination of the four sound transmission loss curves of Figs. 13-2 to 13-5 will show a pronounced dip in the curve in the region

■ Table 13-2 Standard STC contour

Frequency* (Hz)	Sound Transmission Loss, dB
125	24
160	27
200	30
250	33
315	36
400	39
500	40
630	41
800	42
1000	43
1250	44
1600	44
2000	44
2500	44
3150	44
4000	44

*One-third octave intervals.
Ref: ASTM E413-87

of 2500 Hz. We are interested in this dip, because the transmission loss is less where this dip occurs. This means that noise in the general vicinity of 2500 Hz has a better chance of getting through and causing trouble. (An acoustical hole, or at least a thin spot?)

This dip is associated with the bending vibration of the partition. The partition vibrates at its natural frequency of resonance when it is excited by sound energy of that frequency. In Fig. 13-6, the compression wavefronts (solid lines) coincide with the positive peaks of wall vibration, and the rarefaction wavefronts (broken lines) coincide with the negative peaks at this resonance frequency. Due to the vibration of the partition, some sound is radiated on the other side of the partition. Another way of saying this is that the sound transmission loss is decreased at or near this resonance frequency. This is called the "coincidence" effect; the wavelength of the incident sound coincides with the wavelength of the resonance frequency of the partition. The wavelength of 2500-Hz sound is about 0.45 ft.

In addition to the wall's back-and-forth motion, which depends on its weight or mass, a wall can have other motions. Other resonances depend upon a wall's bending stiffness. The practical effect

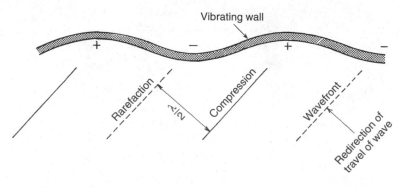

■ **13-6** *An explanation of coincidence dip.*

of all these resonances shows up in the transmission loss curve. The coincidence effect is especially interesting because of its prominence on the transmission loss curves.

The mass law

The mass law gives the average transmission loss for a diffuse source of sound as a function of the wall surface weight and the frequency. The mass law is only a convenient, rough approximation to the performance of single walls. Considering only the mass of a partition, the sound transmission loss increases about 5 dB for each doubling of its surface mass. For a perfectly limp panel (one without any structural stiffness) this figure is 6 dB. A normal partition has some stiffness, hence the 5 dB figure is closer to actuality. In Fig. 13-7, the variation of transmission loss with surface density and frequency is shown.

The surface density is the density (or weight in pounds) of one square foot of partition surface. Transmission loss increases sharply with density and frequency. The 500-Hz line is made heavier to emphasize that it is commonly used to compare partitions as a sort of midscale, midband, rough representation of the entire frequency effect.

Figure 13-8 illustrates a common use of the 500-Hz "mass law." The black dots represent the Sound Transmission Class (STC) ratings of several partitions. Table 13-3 summarizes the data applicable to each of the seven points, pulling together the data from Figs.13-2 to 13-5 and adding three others of a general nature for comparison.

For example, point #1 of Fig. 13-8 shows the STC 38 given by the partition of Fig. 13-2 composed of two leaves (one of 1/2″ and the

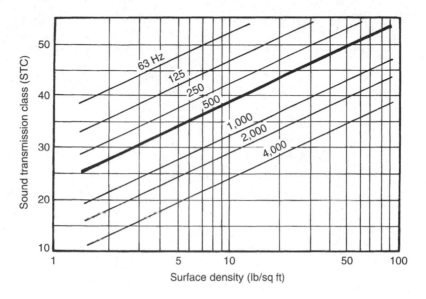

■ **13-7** *STC rating vs. surface density.*

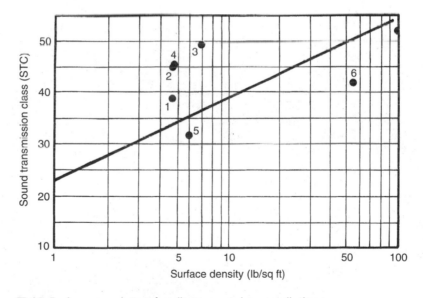

■ **13-8** *A comparison of walls to mass law predictions.*

other of 5/8″ drywall plasterboard), with 3-5/8″ channel and with no glass fiber in the space. Its surface density (4.8 lb/ft^2) alone would give it a 500-Hz transmission loss of only STC 33. The actual value of STC 38 means that the two leaves separated 3-5/8″ perform better than "mass law" with the two leaves stuck together to isolate the mass effect. In other words, the structural form has in-

■ Table 13-3 Sound transmission loss of partitions
(Data Summary for Figure 13-8)

From Figure	Point on Fig. 13-8	Leaf (A)	Leaf (B)	Surface density, lb/sq ft	Leaf spacing	Glass fiber	STC	Source
11-2	1	½"	⅝"	4.8	3-⅝"	—	38	N
11-3	2	½"	⅝"	4.8	2-½"	2"	45	N
11-4	3	2 × ½"	⅝"	7.0	3-⅝"	2-½"	49	N
11-5	4	½"	⅝"	4.8	stagger stud	yes	46	N
	5	½"	½"	4.2	2 × 4 wood studs	—	32	
	6	4-½" brick w. ½" plaster		55.0	—	—	42	
	7	9" brick with ½" plaster		100.0	—	—	52	

N = Northwood 1970

creased the transmission loss about STC 5 dB over that of the mass law alone.

Even though the surface densities of the partitions of Figs. 13-2 and 13-3 are the same (4.8 lb/ft^2), the latter has an STC of 45 while the former is only STC 38. This is in spite of the spacing of the two leaves being decreased from 3-5/8" to 2-1/2". The reason for this improved performance is that 2" of glass fiber has been introduced in the space. Glass fiber does not increase the transmission loss directly; it is far too light in weight for that. It helps by subduing the resonances in the space, which tend to reduce partition performance.

Spot number 3 of Fig. 13-8 indicates still more increase in the STC rating. Surface density has been increased to 7.0 lb/ft^2 by adding gypsum board. Spacing is 3-5/8", and glass fiber has been placed in the space. The result is an increase to STC 49, which is an increase of 15 dB over straight mass law.

Figure 13-5 shows the performance for the traditional staggered-stud wall construction. This reaches STC 46, somewhat short of the STC 49 of Fig. 13-4 in which the surface density was greatly increased.

Spot 5 of Fig. 13-8 is for the simplest wall: 2 × 4 studs with 1/2" drywall attached to each side. The performance is submass law, STC 32, for a surface density of 4.2 lb/ft^2. Spot 2 shows that close to the same surface density can give STC 45 with the addition of a heavier gypsum board sheet and some glass fiber in the space. Spots 6 and 7 of Fig. 13-8 are added to show that the structures considered above can equal or outperform brick walls.

Ways to increase wall insulation

Figure 13-9 illustrates seven simple ways to increase the sound transmission loss of a basic wall structure. Fig. 13-9(A) recalls the need for mass for a wall to be an effective barrier. This mass can easily be added to a wall in the form of sheets of gypsum board by

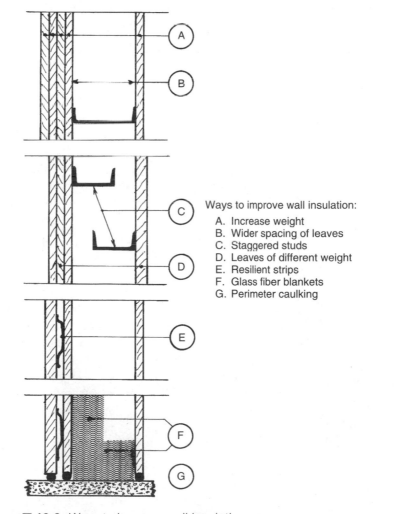

Ways to improve wall insulation:

A. Increase weight
B. Wider spacing of leaves
C. Staggered studs
D. Leaves of different weight
E. Resilient strips
F. Glass fiber blankets
G. Perimeter caulking

■ **13-9** *Ways to improve wall insulation.*

cementing or screwing. Table 13-4 lists the surface density of common sound barrier walls, which vary between 3 and 11 lb/ft^2.

■ **Table 13-4 Surface densities of common partitions**

Leaf A	Leaf B	Surface density, lb/sq ft
Unfilled steel stud partition		
⅜″	⅜″	3.0
½″	½″	4.0
⅝″	⅝″	5.0
⅜″ + ⅜″	⅜″ + ⅜″	6.0
½″ + ½″	½″ + ½″	7.5
⅝″ + ⅝″	⅝″ + ⅝″	10.0
Unfilled wood stud partition		
⅜″	⅜″	4.0
½″	½″	5.0
⅝″	⅝″	6.0
⅜″ + ⅜″	⅜″ + ⅜″	7.0
½″ + ½″	½″ + ½″	8.5
⅝″ + ⅝″	⅝″ + ⅝″	11.0

Note: The densities above include weight of studs.

If the spacing between the two leaves of a wall is eliminated or made very small, the transmission loss will be reduced to straight mass functioning as though it were a single-leaf wall of combined weight. For two leaves spaced widely apart, the transmission loss approaches the sum of the two as though two separate walls existed. Our practical wall spacing of about four inches is somewhere between the two extremes. The transmission loss is increased only a small amount by increasing the spacing from four to six inches, but this is one factor in the significant increase for staggered-stud walls.

Staggered-stud walls [shown in Fig. 13-9(C)], are normally built on 6″ plates instead of the usual 4″ plate. Though nominal in itself, the added spacing gives some increase in transmission loss. Making the mass of the two leaves different [as in Fig. 13-9(D)] makes the resonance frequencies of the two leaves different. When these frequencies coincide, the dip in the transmission loss curve is magnified at that frequency. Staggering the resonance frequencies by making the leaf mass different tends to smooth the transmission loss curve.

Figure 13-9(E) illustrates the use of resilient strips for the purpose of providing some isolation of a layer of gypsum board from the wall itself. This also acts in the direction of making the resonance points of the two leaves appear at different frequencies, avoiding deep dips in the transmission loss curve.

The use of thermofiber (building insulation) in the space between the two leaves of a wall [Fig. 13-9(F)], is an easy way to increase the transmission loss of a wall about 5 dB. This glass fiber tends to control various cavity resonances, which tend to decrease transmission loss.

Figure 13-9(G) places emphasis on the importance of sealing around all edges of a partition. The importance of liberal use of sealant cannot be overstressed. It is the inexpensive way of achieving added transmission loss in a wall. Normal framing always results in cracks, which might not be important in nonacoustical walls, but which are very important in walls associated with studios and other sound-sensitive spaces. The cracks present in normal framing are always there, and the sound penetrating cracks under, over, and around a wall can easily mean the loss of 10 dB of STC rating. Figure 13-10 illustrates both the problem and the solution.

For example, before the 2×4 or 2×6 plate is bedded on the floor, several strips of acoustical sealant should be applied to the floor

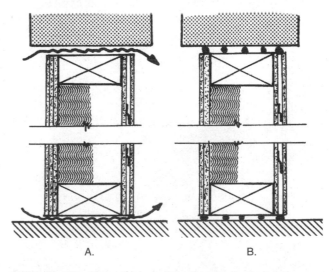

■ **13-10** *The importance of sealing wall elements.*

and/or the plate to seal the ever-present crack between the concrete and the plate. Every edge-framing member should be so sealed. Additional sealant should be applied to the periphery of the gypsum board wall. The sealant used should be of the non-hardening variety, often referred to as "acoustical sealant."

Strengthening an existing wall

It is often necessary to adapt an existing space for a studio or work room. Existing walls are rarely built for high transmission loss. A weak existing wall can be strengthened by constructing another wall in front of it, but the construction of that wall will determine whether it is a success or failure as a sound barrier. Figure 13-11 details what might be called an optimum, practical approach to such a wall.

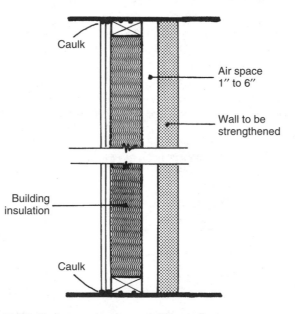

Caulk

Air space
1″ to 6″

Wall to be
strengthened

Building
insulation

Caulk

■ **13-11** *Improving an existing wall.*

The air space is important; one inch is a minimum, but use more if floor space allows. From there on, the wall simply embodies the features that have been described, such as thermofiber (building insulation) in the airspace and multiple layers of gypsum board. The outer layer of gypsum board could be mounted on metal resilient strips, and more than two layers of gypsum board could be included to squeeze out that last couple of dB of STC. Such extreme steps are probably unwarranted because of flanking sound traveling around the barrier.

Flanking sound

If placing an ear on the rail makes it possible to hear an oncoming train that is several miles away, why is it surprising that solid materials are good carriers of sound?

In the modern high-rise concrete structure, elevator sounds and other noises are carried throughout the structure with very little loss. Wooden structures are also good conductors of sound. Table 13-5 compares the attenuation of sound in iron, brickwork, concrete, and wood. In concrete, for instance, sound can travel 100 ft with an attenuation of only 1 to 6 dB.

■ Table 13-5 Attenuation of longitudinal
waves (Harris, 1957)

Material	Attenuation dB/100 ft
Iron	0.3–1
Brickwork	0.5–4
Concrete	1.0–6
Wood	1.5–10

Structure carries sound from one point to another, but radiation of that sound into the hopefully quiet area is accomplished largely by walls and floors acting as diaphragms. Instead of the noise staying in the beams and columns of the structure, it sets the walls and floors to vibrating, radiating efficiently into the air within the rooms. In this way, protection from both airborne and structure-borne sound must be considered.

Figure 13-12 illustrates the travel of noise through structure to walls which radiate into the space of the protected area. The specific case of a frame structure is shown in Fig. 13-13. Flanking sound can travel by a combination of air- and structure-borne paths from the noisy space to the protected area though an attic space or a subfloor space. Such flanking sound can nullify careful work spent on strengthening a shared wall.

Masonry walls as sound barriers

The performance of brick and concrete block walls are summarized in Table 13-6. The STC of brick walls can be equaled by

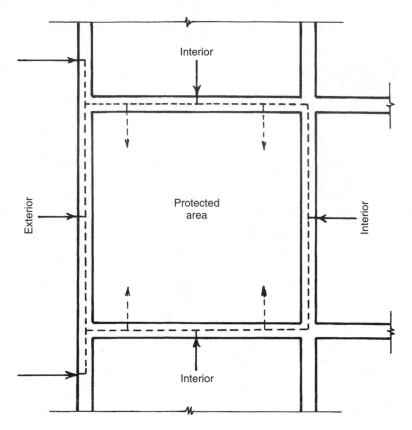

Interior

Exterior

Protected
area

Interior

Interior

■ **13-12** *Structure-borne sound effects.*

frame walls, but plastered concrete block walls, with their STCs of 57 and 59 are difficult to reach with frame construction. Circumstances often dictate which walls are most practical, both from the acoustical and the construction viewpoints.

Figure 13-14 shows the very smooth measured transmission loss curves of both unplastered single- and double-leaf concrete block walls. In comparison with the concrete block walls of Table 13-6, it would appear that plastering is more effective on the single-leaf wall (an increase of 6 dB) than the double-leaf wall (no increase).

Doors as sound barriers

Common doors are probably the weakest link of all the sound barriers in budget studio construction. Hollow-core household doors have STCs in the low 20s, and solid-core household doors have STCs in the upper 20s. Even these values assume good weather-stripping.

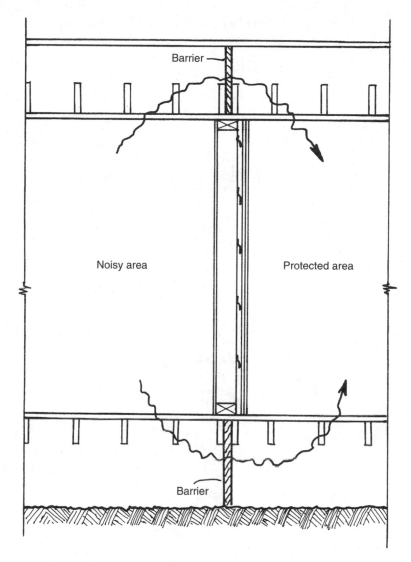

■ 13-13 *Flanking paths and barriers.*

The only truly satisfactory doors for sound-sensitive rooms are those metal doors and jambs designed especially for the task and supplied by the industry. The Overly Model STC488861 door is shown in Fig. 13-15. It is 1-3/4″ thick and rated as STC-48, with a surface density of 9.9 lb/ft².

Summary

Figure 13-16 summarizes many of the points made in this chapter in regard to possible ways the sound transmission loss of a wall

■ Table 13-6 Masonry sound barriers summary

Wall description	Surface density, lb/cu ft	STC	Ref.
Brick wall, 4-½″ plastered both sides	55	42	Table 13-3
Brick wall, 9″ plastered both sides	100	52	Table 13-3
Double 6″ concrete block walls spaced 6″	100	59	Fig. 13-14
Single 12″ concrete block wall	100	51	Fig. 13-14

might be increased economically. Mass is the most vital element of any noise barrier. In Fig. 13-16(A) the mass of a partition is low, but it can be increased easily by adding multiple layers of gypsum-board panels. These can be screwed or cemented.

In Fig. 13-16(B) an increase in spacing of the two leaves of a partition is achieved by staggered studs. This will increase its STC

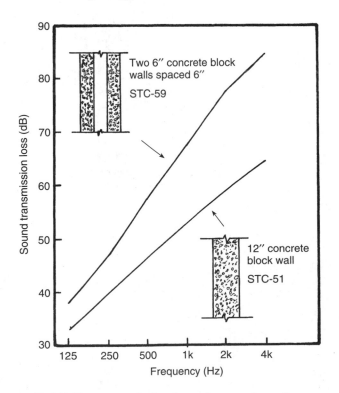

■ **13-14** *The transmission loss of concrete walls (Egan 1972).*

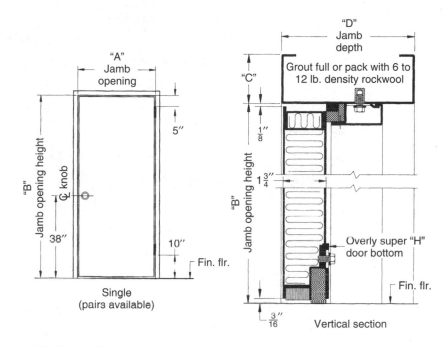

A
Jamb opening

B
Jamb opening height

Ⅼ knob

38″

5″

10″

Fin. flr.

Single
(pairs available)

D
Jamb depth

Grout full or pack with 6 to
12 lb. density rockwool

C

B
Jamb opening height

$\frac{1}{8}$″

$1\frac{3}{4}$″

Overly super "H"
door bottom

Fin. flr.

$\frac{3}{16}$″

Vertical section

■ **13-15** *An Overly door.*

value. In (C), staggered studs of metal or wood are used to increase the spacing between the two leaves. In addition, another layer of gypsum board has been added to one side to make the surface densities of the two sides different. This places their resonances at different frequencies, smoothing the transmission loss of the wall. In (D) the surface density of the two leaves is made different by having layers of wall board of different thicknesses on each side to vary the surface densities. In (E) the use of resilient strips is suggested. If one face of a wall is mounted resiliently to the structure, the STC can be increased significantly. (F) reminds us that a small improvement can be expected by using glass fiber in the space, or by increasing its thickness. The improvement by adding glass fiber is modest, but is inexpensive and worth the effort. (G) represents a most important factor which can have a major effect in increasing the transmission loss of a wall. This is to use two completely separate walls. In all arrangements, proper sealing is most important. Copious use of tubes of caulking material could be the most important single step in the entire process.

The burden of this chapter is to examine the options available to the studio designer and builder with regard to barriers to protect the sound-sensitive space from the encroachment of noise.

Summary

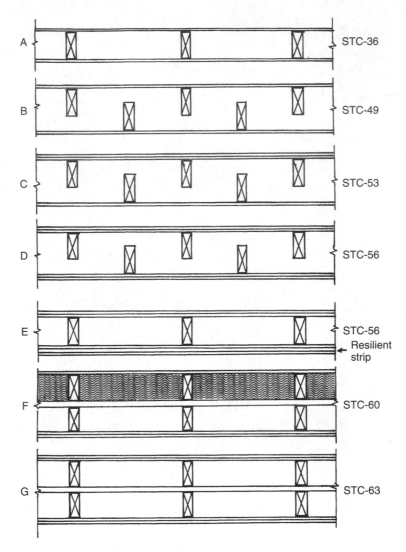

■ **13-16** *The STC ratings for many wall forms.*

Actually measured transmission loss curves of Figs. 13-2 to 13-5 have been examined. Such measurements require special facilities and experienced operators and are impractical for specific structures for all but the largest jobs.

The STC shorthand, which attempts to give a practical and simple shortcut, has been widely accepted, but its limitations must always be kept in mind. The ultimate test of any noise barrier system, which can be made only when the studio is completed, is noise level measurements within the studio itself.

Floor/ceiling construction and performance 14

FOOTSTEP SOUNDS HAVE LONG BEEN A SOURCE OF WOE TO builders, residents, and studio operators alike. In addition to such noises ruining recordings, the sounds associated with the activity of people in the space above has always been a source of complaints from the people living below. Such noises destroy the concept of privacy in living quarters, and can also be a source of much trouble when the noises penetrate a sound-sensitive space such as a recording studio. The following living-space example brings a better understanding of the problems in studios.

Data on the footfall noise problem

Warren E. Blazier, Jr., consultant in acoustics, and Russell B. Dupree of the Office of Noise Control, State of California, have given an account of how important footfall noise can be (Blazier 1994). Recently the homeowners in a large "luxury" condominium complex in the San Francisco area brought an $80 million class-action suit against the developers. The major claim was based on annoyance caused by footstep noise generated by the activity of residents above being transmitted through the structure to their neighbors below. Even when cooperative upstairs neighbors agreed to go barefooted or to wear soft-soled shoes, the "thuds" and "thumps" were still painfully audible below. The footfall impacts also resulted in *feelable* vibrations of the floor/ceiling structure, vibrations that even affected closet doors and light fixtures. Complaints were general throughout the building.

The average purchase price of the apartments was about $750,000, with marketing claims of "luxury" acoustical privacy. Because of the upscale nature of the project, the builders incorporated special design features to provide significantly better impact noise insulation than that required by California construction standards for multifamily housing. The floor/ceiling construction in question is shown in the section of Fig. 14-1. Double joists were used to stiffen the floor plane. The ceiling of double 1/2″ gypsum board is mounted on resilient channels. The floor above starts with the usual 3/4″ plywood subfloor, on which is placed Kraft paper, a resilient mat, reinforcing wire, and 1-1/4″ mortar. On top of this is 3/8″ ceramic tile. To minimizing flanking, a dense glass fiber perimeter strip isolates the floor from the structure (a floating floor, in essence).

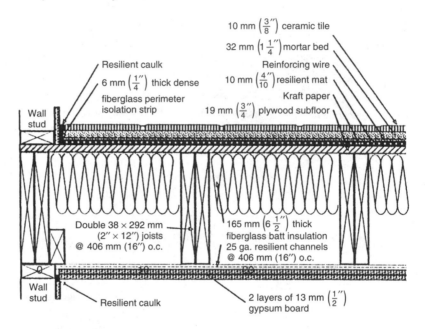

■ **14-1** *Blazier floor/ceiling construction.*

To provide ammunition for the defense, it was decided to build an off-site laboratory mockup duplicating a pair of typical stacked rooms. The floor/ceiling structure was that described in Fig. 14-1. The lower test room was used to measure impact sound pressure levels resulting from three types of noise sources: a standard impact tapping machine, a standardized "live-walker," and a calibrated tire drop. The measurement microphone on a 20-inch boom was located in the center of the room below, and was rotated slowly in a horizontal plane during the integrating period. Data was obtained in one-third octave bands from 2.0 Hz to 4.0 kHz.

Among the many conclusions, the one concerning footwear is very interesting. In the "live-walker" tests it was found that for frequencies below about 63 Hz, the impact sound pressure levels below the walking surface were amazingly close together for leather heel/leather sole, rubber heel/leather sole, track shoe, and barefoot cases. Figure 14-2 shows that the peak energy of these live-walker tests fall in the 15–30-Hz region. This coincides with the fundamental natural frequency of the floor/ceiling system which, with typical lightweight structural framing, is also between 15 and 30 Hz.

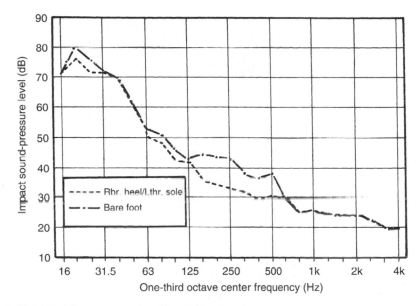

■ **14-2** *Measurements of footfall noise.*

The addition of floating floors or carpeting decreases the transmission of the higher-frequency components of footfall noise, but there is no economically practical method of avoiding the thuds and thumps of footfalls with typical lightweight structural framing. To obtain the stiffness necessary to reduce the thuds and thumps, a concrete structural floor system is required.

In Fig. 14-2, note carefully all the extensive measurements made three octaves below the usual low-frequency limit of 125 Hz. Going no lower in frequency than 125 Hz misses common impact sounds 40 dB higher than at 125 Hz. Carpet is effective in reducing noises above 125 Hz, but completely ineffective in the 15–30-Hz range.

Noise in the frequency range below 30 Hz is becoming more important as audio systems, including auxiliary bass loudspeaker elements, radiate sound efficiently in this region. Perhaps footfall noises from the people upstairs can be used in demonstration discs, along with trains running through the listening area to dramatize the effectiveness of the system!

Frame buildings

The first and most distressing factor in floor/ceiling problems is that usually there is *no access to the floor side*. In other words, the only hope for improving the shielding of our studio from the neighbors' noise from above is what can be done to the ceiling. Floor/ceiling constructions in frame structures are quite limited in scope, and most of them would be included in the three shown in Fig. 14-3.

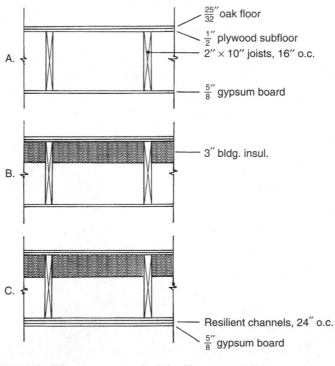

■ **14-3** *Three common floor/ceiling constructions.*

In Fig. 14-3(A), the 2″ × 10″ joists commonly have a 5/8″ gypsum board ceiling nailed to the lower edges and a 1/2″ plywood subfloor above. The finish flooring above is probably something very much like the 25/32″ tongue-and-groove oak flooring pictured.

This is the "as found" condition; what can be done about it without access to the floor side?

The first step would be to remove the gypsum board ceiling fastened to the lower edge of the joists and place 3" building insulation between the joists. This insulation could be 6" or 10" thick, but very little would be gained by going beyond the 3" batts.

Something very worthwhile can be done to the ceiling itself by mounting resilient strips on the joists and a layer of gypsum board to the strips, and then adding a second layer of gypsum board. The resilient strips could instead be mounted between the two layers of gypsum board. We are still limited by not having access to the floor side above. What performance can be expected from these corrections made below?

The measured transmission loss curves for the three conditions of Fig. 14-3 are shown in Fig. 14-4. It is noted that the curves are quite smooth and free from untoward resonance effects. The as-found condition (A) gives an STC-37, and adding the thermofiber (B) gained only 3 dB to STC-40. Making a decent ceiling (C) by two layers of gypsum board with one of them resiliently mounted brought it to STC-47. Is achieving a 10-dB improvement to STC-47 all that can be done from below? There is still hope; read on.

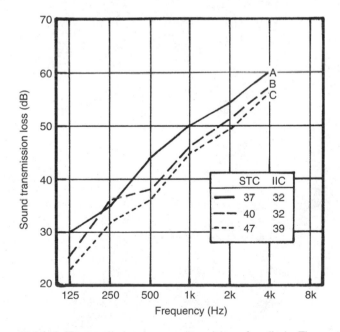

■ **14-4** *Transmission loss comparison of walls in Fig. 14-3 (Egan 1972).*

Resilient hangers

So far, only the resilient channels used to provide some insulation to gypsum board sheets have been mentioned. Another system that can replace or add to the resilient channels is illustrated in principle in Fig. 14-5. If a bit of ceiling height can be sacrificed, a suspended ceiling can be installed, one that is isolated from the structure by resilient hangers. The resilience of the hanger illustrated in Fig. 14-5 is provided by the shaded material, which could be compressed glass fiber of the proper density. The load of the ceiling gypsum boards and supporting frame must be distributed between hangers to keep the deflection of each hanger within its rated range.

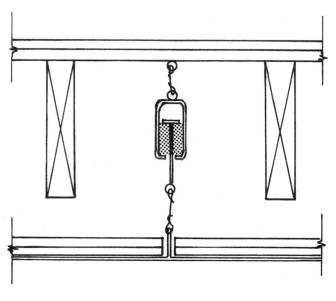

■ **14-5** *A resilient hanger.*

There are several hanger designs available. The two in Fig. 14-6 utilize neoprene alone or a combination of neoprene and steel spring. The frequency range of these two products differ; the selection of the proper hanger for the specific job is important, and data about each is available from the manufacturer. The weight of the ceiling must be carefully matched to the deflections of the hangers to achieve maximum insulation. The periphery of the suspended ceiling should be sealed with a nonhardening acoustical sealant.

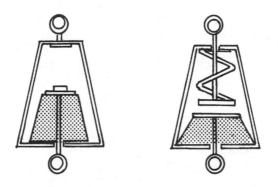

■ **14-6** *More forms of resilient hangers.*

Floating floors

It may be possible to make friends with the neighbors above with the hope that something can be done to their floor to control their noise, but do not expect too much in controlling the actions of other people. With the cooperation of the persons above, however, a floating floor could possibly be installed.

Floating floors normally are not a "budget" subject, related as they usually are to concrete structures and world-class studios. A sound-sensitive room in a frame structure might be more than a dream if access can somehow be obtained to the floor above. With such access the budget floating floor of Fig. 14-7 might have limited application.

193

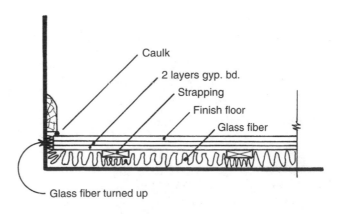

Caulk

2 layers gyp. bd.

Strapping

Finish floor

Glass fiber

Glass fiber turned up

■ **14-7** *A budget floating floor.*

The weight of the two layers of gypsum board and the finish floor is supported by a layer of glass fiber. The springiness of the partially compressed glass fiber is the secret of success. If it is totally compressed, the plywood/finish floor layers would essentially rest on the basic floor, and little insulation would result. A basic principle is that the gypsum board/finish floor mass must not touch the structure, including the edges. In Fig. 14-7 the glass fiber turned up at the edges is used to isolate the floor mass edge from the structure. A strip of compressed glass fiber could also be used to provide insulation from the wall.

Floor/ceiling structures and their performance

Figure 14-4 gives the transmission loss performance of the three floor/ceiling arrangements of Fig. 14-3. Just by adding the thermofiber the STC increased the usual 3 dB from 37 to 40. Adding the resilient channels, however, increased the STC from 40 to 47, a gain of 7 dB. From the bare-bones "as found" situation of Fig. 14-4(A) a total of 10 dB in STC rating has been gained by the addition of resilient channels and glass fiber. This is close to the maximum improvement that can be expected, unless some new principles are introduced.

Figure 14-4 includes IIC measurements as well as STC. IIC stands for Impact Isolation Class, and is a single-number rating of the impact sound performance of floor/ceiling constructions over a standard frequency range. This impulse single-number rating is comparable to the STC single-number rating for steady-state sounds. The higher the IIT rating, the more efficient the floor/ceiling construction in reducing impact transmission of sounds such as footfalls.

Attenuation by concrete layers

Figure 14-8 gives some insight into the contribution to attenuation that might be expected by adding a 1-1/2" layer of troweled cellular concrete of density 105 to 120 lb/ft^3.

The two floor/ceiling constructions of Fig. 14-8 (Grantham and Heebink, 1973) are identical, except one has 3-1/2" of glass fiber and resilient channels and the other does not. An STC-59 is attained by the construction having the glass fiber and the resilient channels, a very worthwhile increase of 10 dB. This 10-dB increase in STC rating reminds us of another 10-dB increase, that from STC-37 to STC-47 of Fig. 14-4 obtained by adding some gypsum

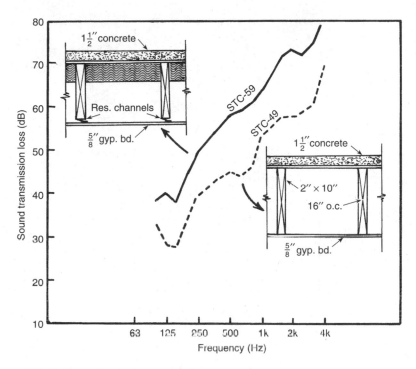

■ 14-8 *The effect of concrete topping.*

board, glass fiber, and resilient channels to the bare-bones "as found" situation of Fig. 14-3(A).

By adding the concrete topping, more gypsum-board layers, glass fiber, and resilient channels, an increase from a poor floor/ceiling structure offering STC-37 to a very respectable STC-59 situation results. This is a total of 22 STC points available for use.

Figure 14-9 is a convenient comparison of the best floor/ceiling structure without a concrete layer with the best floor/ceiling structure with a concrete layer. In other words, an estimate of the effect of the concrete layer at different frequencies can be made by comparing the STC-59 curve of Fig. 14-9 with the STC-47 curve below it. The concrete has contributed about 10 dB at low frequencies, about 15 dB at 500 Hz, and about 20 dB above 2 kHz.

Plywood web versus solid-wood joists

Some floor/ceiling systems in frame structures use plywood web beams instead of conventional 2″ × 10″ or 2″ × 12″ wood joists. They are made with 2″ × 3″ flanges and a 3/8″ plywood web. These ply-

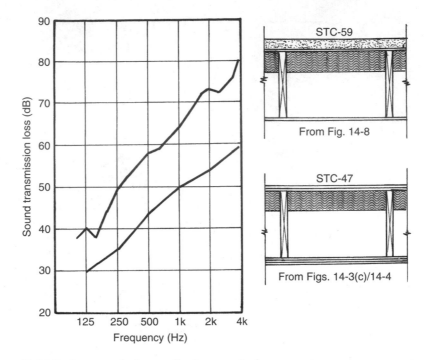

■ 14-9 *A concrete layer effect separated.*

wood web beams, illustrated in Fig. 14-10, offer certain advantages. As they are usually fabricated off the building site, they contribute to speed and efficiency. From the viewpoint of the architect or designer, 12″ plywood web beams allow greater spans. From the viewpoint of the acoustician, they contribute to the stiffness of the floor plane and thus higher transmission loss to low-frequency noises.

Figure 14-11 presents the transmission loss characteristics of two good floor/ceiling systems, both offering an STC-58 rating (Grantham and Heebink, 1973). Both use resilient channels, both have 3-1/2″ glass fiber, and both have 3/4″ plywood subfloor. They differ only in that the floor/ceiling represented by the solid curve uses 12″ plywood web joists, while the other (the broken curve) uses 2″ × 10″ joists. There is little consistent difference between the two curves. The plywood web joists give 5-dB-higher attenuation to sound at 100 Hz than the 2″ × 10″ joists. The stiffening of the floor plane by the plywood web joists results in greater low-frequency attenuation, which is just beginning to be noticeable in Fig. 14-11. The reverse is true in the 1000–2000-Hz region, where the wood joists offer greater attenuation than the plywood web joists. Carpets help in the 1000 to 2000-Hz region, but not in the 15–31-Hz region.

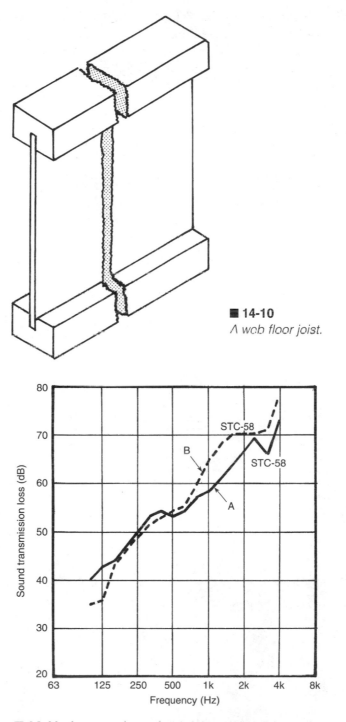

■ **14-10**
A web floor joist.

■ **14-11** *A comparison of web joists with solid-wood joists.*

It is instructive to note in Fig. 14-2 the low-frequency rise of impact noise level of some 40 dB between 125 Hz and 16 Hz. There is much footfall and other noise at low frequencies that is undetected by thinking patterns and measuring systems confined to the old 125–4000-Hz region.

Observation window

15

IT IS IMPORTANT TO BE ABLE TO SEE WHAT IS GOING ON IN the other room. For example, those in the control room of a studio must know exactly what is happening in the studio. The reverse is also true. The transparency of glass makes it the first product that comes to mind to establish visual communication between two rooms. The first lesson is that if a glass window is set in a high-transmission-loss wall, the transmission loss of the glass becomes important.

Glass is so common that it is easy to think that everything about it is known. Can a substance that is brittle, without a yield point, and that fractures easily be safely handled? Yes, with care. There are many kinds of glass. Over 98% of the glass produced in the United States is made by the *float glass* process. The molten glass is poured continuously from a furnace onto a large bed of molten tin. The molten glass literally floats on the surface of the tin, slowly solidifying as it travels over the tin. After several hundred feet of travel through a lehr, it emerges as a continuous layer of glass at approximately room temperature. The glass now is flat, fire finished, and with virtually parallel surfaces.

Sheet glass accounts for a very small proportion of U.S. production. Some imported sheet glass is used in thicknesses of 1/8th inch or less. *Plate glass* is manufactured by the grinding and polishing process, but is no longer produced in the United States. It has been replaced by float glass. *Rolled glass* is fed from the furnace to rollers, which produce the desired thickness. The three general types of rolled glass are: figured/patterned, wired, and art/opalescent glass. Forgetting all these other types of glass, *float glass* provides adequate transparency and low distortion for use in observation windows for studios and other sound-sensitive rooms.

The observation window

If the idea of picture windows and extensive skylights can be successfully outvoted in the early planning stages, about the only glass left in the studio is the observation window and possibly a few peepholes in doors. The classic observation window is the one between the control room and the studio. Sometimes it is important to allow visitors to peer into the busy studio or workroom, which would require an observation window between the studio and an exterior hall.

Some studios located in highly scenic spots have incorporated a view window to impress the talent and clients. Every such extra window makes it that much more difficult to achieve sufficiently low background noise levels, and demand is growing for lower noise in recordings.

Glass as a sound barrier (single leaf)

The single-leaf or single-pane window, such as the common household window, is a sound blocker of sorts. Those living near an airport, however, find that windows of this type allow too much aircraft noise to pass through. It is the mass of the glass that controls the passage of sound through it according to the expression:

$$TL = 20 \log (fm) - 48 \qquad (15\text{-}1)$$

in which

TL = transmission loss, dB
f = frequency, Hz
m = surface density, lb/ft^2

Taking the density of glass as 160 lb/cu ft, the surface density may readily be found for any thickness of the glass pane. There is some lack of unanimity regarding the value of the constant to be subtracted. Quirt (1982) recommends 48 dB, but others prefer 34 dB. This is of oblique importance in this discussion, because the transmission loss of various glass arrangements to be presented has been determined by actual measurements. The mass law, expressed graphically in Fig. 15-1, emphasizes that mass is the major component of transmission loss of glass, although glass configuration (multiple panes, differing thickness, and subduing of resonances) will be presented as tricks to enhance the transmission loss of an observation window.

The double-glass window

If a single pane offers insufficient transmission loss, the logical observation is that more panes will offer more. There are many

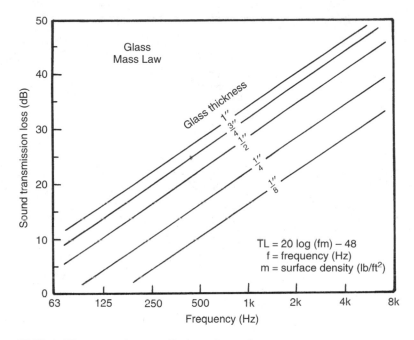

Glass
Mass Law

Glass thickness

1"
3/4"
1/2"
1/4"
1/8"

TL = 20 log (fm) − 48
f = frequency (Hz)
m = surface density (lb/ft²)

■ **15-1** *The mass law applied to glass plates.*

qualifications to this observation, but the transmission loss of the double-glazed windows of Fig. 15-2 (Sabinc 1975) shows a definite increase in transmission loss over the single-pane mass law of Fig. 15-1. In Fig. 15-2, two double-glass windows with approximately 4-inch spacing are compared on the basis of thickness of glass.

The solid curve (Libby-Owens-Ford) represents measurements made with a 1/8″ pane and a 3/32″ pane. The broken-line curve (National Bureau of Standards 1975) shows measurements made on one pane of 1/2″ and a second of 1/4″. Later the reasons for using glass of dissimilar thicknesses will be discussed in detail. The thinner glass window and the heavier glass window have similar transmission loss in the 1–3 kHz frequency region, but the heavier glass is far superior below 1 kHz. The STC (Sound Transmission Class) ratings, which show only an advantage of 4 dB for the heavier glass, are quite inadequate for describing the performance of these two windows over the audible band.

In Fig. 15-3 the measured transmission loss of another double-glass window is shown (Quirt 1983). In this window two 1/4″ glass panes are separated 6″. The overall transmission loss of the 6″ spacing is quite similar to that of the heavier glass window of Fig. 15-2. The only way the modest specific effects of dissimilar panes, glass surface density, and glass spacing can be observed is in di-

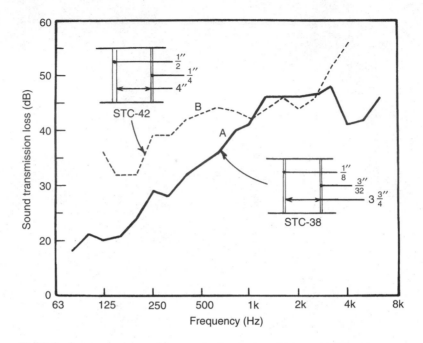

■ **15-2** *A comparison of transmission losses in two double-glass windows.*

rect comparisons such as will be made later. The irregularities of the usual measured-transmission-loss curves, due principally to resonances, tend to hide these other variables of direct interest.

Acoustical holes

An "acoustical hole" is a phenomenological name for a sliver of the audible spectrum in which sound more readily passes through a wall or observation window. Unlike a knothole, an acoustical hole is never transparent, but always somewhat opaque to sound. In other words, sound within a narrow frequency range is attenuated a few dB less than at other frequencies. It shows up on the transmission-loss curve as a dip at the frequency of the hole. Acoustical holes are usually traceable to resonances; numerous examples follow.

Several different resonance effects alter the shape (and effectiveness) of the transmission-loss curve of any glass window arrangement. The mass-air-mass resonance of a double window is the result of the mass of one glass plate being coupled to the other glass plate by the air of the cavity between them. Sound impinging on the first glass plate will set the plate to vibrating. The air in the

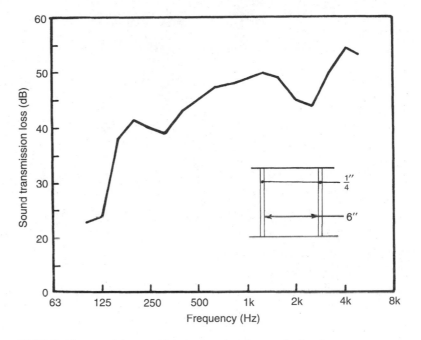

■ 15-3 *Heavy glass and large spacing transmission loss.*

cavity acts like a spring, setting the second glass plate to vibrating. This resonant system can be likened to a mass attached to each end of a spring.

Quirt made a study of the effect of glass spacing on two 1/8″ glass panes (Quirt 1982). He found that at 250 Hz the sound transmission-loss curve went through a very pronounced null at spacings between one-half inch and one inch, as shown in Fig. 15-4. At 800 Hz, the sound transmission loss showed no such dip. Mass-air-mass resonance is largely a low-frequency effect; in fact, with certain glass panes at certain spacings, the resonance frequency is so low that it does not appear in the usual 63 Hz to 8 kHz measuring range.

The mass-air-mass resonance frequency can readily be estimated by the following approximate equation:

$$f = 170 \sqrt{\frac{m_1 + m_2}{m_1 \, m_2 \, d}} \qquad (15\text{-}2)$$

in which

f = mass-air-mass resonance frequency, Hz
m_1 = surface density of glass A, lb/ft^2
m_2 = surface density of glass B, lb/ft^2
d = spacing between glasses, inches

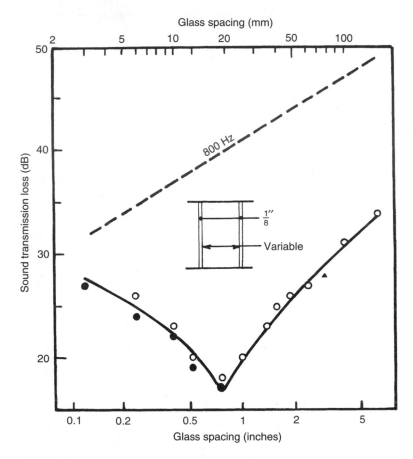

■ **15-4** *The mass-air-mass resonance effect.*

The mass-air-mass resonance frequency has been calculated from Eq.15-2 for many glass weights and many spacings and plotted in Fig. 15-5. This illustrates that only for the lighter glass and smaller spacings are the resonance frequencies above 100 Hz. Mass-air-mass resonance is primarily a low-frequency effect.

A prominent mass-air-mass resonance appears in the sound transmission-loss curve of Fig. 15-6 measured by Quirt (Quirt 1982). The notch at about 400 Hz is what has been called an acoustical hole in the glass. Sound is attenuated about 5 dB less near 400 Hz than at adjoining frequencies. Both the very small spacing of 1/4″ and the very light glass of 1/8″ thickness have purposely been selected to bring the mass-air-mass resonance up to about 400 Hz to demonstrate its existence. For more practical double-glass windows, this resonance might be at too low a frequency to appear on the measured transmission-loss curve.

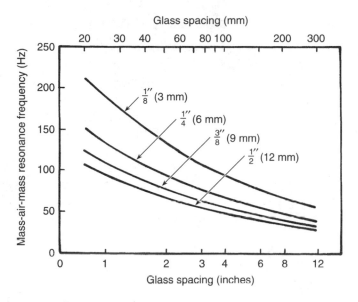

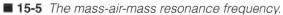

■ **15-5** *The mass-air-mass resonance frequency.*

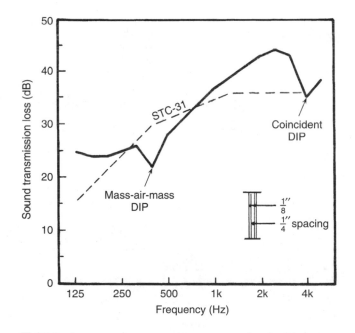

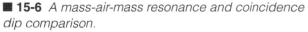

■ **15-6** *A mass-air-mass resonance and coincidence dip comparison.*

Acoustical holes: coincidence resonance

The other dip in Fig. 15-6 at about 4 kHz is called a coincidence dip. It, too, can be classed as an acoustical hole, but is caused by an entirely different process. The flexural, bending vibrations of the glass panel interact with the impinging sound in such a way that an abnormal amount of sound is transmitted through the glass at the coincident frequency. When the phase of the pressure crests of the incident sound coincides with the vibrational crests of the panel, this coincidence effect results in lowering of the sound transmission loss. This means that sound at or near this frequency penetrates the window more easily. The frequency at which coincidence occurs may be estimated from:

$$f = 500/t \qquad\qquad (15\text{-}3)$$

in which

f = coincident frequency, Hz
t = thickness of glass, inches

The window of Fig. 15-6 employs glass 1/8″ thick. Equation 15-3 then becomes $t = (500)(8) = 4000$ Hz, which is close to the observed frequency of the coincident dip.

Acoustical holes: standing waves in the cavity

The space between the two panes in a double-glass window is capable of supporting standing waves, much as sound in a studio or listening room. In Fig. 15-7 the three normal modes are those associated with the length, height, and depth of the cavity. In addition there are the tangential modes striking four walls, and the oblique modes striking all six surfaces. Because of the greater number of reflections encountered in a single circuit, the tangential modes and the oblique modes have lower energy levels (-3 dB and -6 dB below the axial modes) as pointed out by Morse and Bolt in their classic paper (Morse 1944). The frequencies of all three modes may be calculated by Equation 12-1.

Table 15-1 lists the frequencies associated with a window cavity having dimensions of $8 \times 4 \times 0.5$ ft. The lowest axial-mode frequency associated with the 8-ft length (the 1,0,0 mode) is 70.6 Hz. The lowest axial-mode frequency associated with the 4-ft width of the cavity (the 0,1,0 mode) is 141.3 Hz. The lowest axial-mode frequency associated with the 6″ depth of the cavity (the 0,0,1 mode) is 1130 Hz.

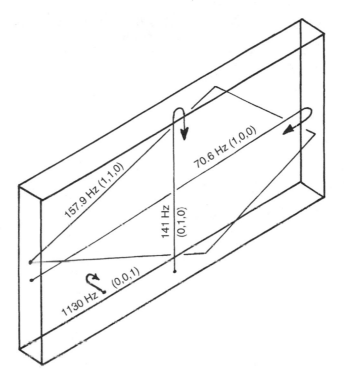

■ 15-7 *Resonances in the space between two glass panes.*

It is important to know these frequencies in order to determine whether the absorbent material on the periphery of the space between the glass plates (the "reveals") is capable of damping these modal resonances sufficiently. If not damped, there is the possibility of minor acoustical holes appearing at certain frequencies. The thickness of this periphery absorbent is severely limited by the space available. The 70.6- and 141.3-Hz axial modes are the lowest to be absorbed. The 1-inch-thick glass fiber commonly used as cavity absorbent does not absorb these frequencies well, and there is no space for a Helmholtz resonator absorber. Conclusion: The cavity resonances at 70 and 141 Hz will probably be close to full strength in the finished window.

Effect of glass mass

Although Fig. 15-8 is primarily a study of the effect of spacing between the glass panes and the glass mass, it contains information on the effect of mass. The upper curve is for two 1/4" glass plates, while the lower one is for two 1/8" glass plates. This doubling of glass weight results in an increase of 3 dB on the STC scale.

■ Table 15-1 Observation window cavity resonances
(cavity dimensions: 8 ft × 4 ft × 0.5 ft)

p q r	Axial	Tangential	Oblique
1, 0, 0	70.6 Hz		
0, 1, 0	141.3		
1, 1, 0		157.9 Hz	
0, 0, 1	1130.0		
1, 0, 1		1132.0	
0, 1, 1		1138.0	
2, 0, 0	141.3		
2, 0, 1		1138.0	
1, 1, 1			1147.0 Hz
0, 2, 0	282.5		
2, 1, 0		199.8	
1, 2, 0		291.2	
0, 2, 1		1164.0	
0, 1, 2		2264.0	
2, 1, 1			1147.0
1, 2, 1			1147.0
2, 2, 0		315.8	
3, 0, 0	211.9		
0, 0, 2	2260.0		
3, 1, 0		254.6	
0, 3, 0	423.8		
2, 2, 1			1173.0
3, 0, 1		1149.0	

Effect of spacing of glass

Figure 15-8 is the result of a great many careful measurements of double-glass windows (Quirt 1983) for 1/4″ and 1/8″ glass. As mentioned, the slope of the two curves indicate that there is a gain of 3 dB in STC rating for every doubling of the mass of the glass. The curves also show that doubling the spacing results in approximately a 3 dB gain in STC rating. In the search for higher transmission loss with double-glass windows, several modest gains must be combined to get the best window.

Effect of dissimilar panes

Figure 15-2 compares two double-glass windows, both with approximately 4-inch spacing, one with 1/8″ and 3/32″ glass and the other with 1/2″ and 1/4″ glass (Sabine 1975). As both examples use dissimilar glass, specific evaluation of the dissimilarity factor is

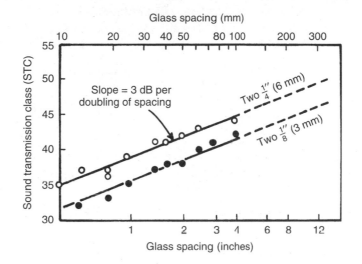

■ **15-8** *The effect of spacing on STC rating.*

not possible. However, it is known that glasses of different mass on the two sides will distribute the mass air-mass resonances, resulting in a smoother transmission-loss curve.

The mass-air-mass resonance of the lighter window (the lower curve of Fig. 13-2) was found by calculation to be about 90 Hz and that of the heavier window (the upper curve) 57 Hz. These are not discernible on the two curves. The coincidence effect involves only a single pane of glass. By calculation, the coincidence frequencies of the lighter window were found to be 4000 Hz for the 1/8″ glass and 5300 Hz for the 3/32″ glass. There is a pronounced dip in the heavy curve in this region of the spectrum. The coincidence frequency for the 1/4″ glass was found to be 2000 Hz, and that of the 1/2″ glass was 1000 Hz. Small dips can be seen in the broken-line curve at both of these frequencies. Staggering these dips by using glass of different thickness tends to smooth the transmission-loss curve and minimize the acoustical hole effect. If both leaves resonate at the same frequency, the dip would have been deeper.

Effect of laminated glass

Glass may be laminated by placing a layer of polyvinyl butyral, often of 0.015″ thickness, between two layers of glass. Architectural laminated glass, the most familiar form, consists of two plies of glass bonded together by a plastic interlayer (usually PVB) under a pressure of about 250 psi and a temperature of 250–300 degrees

Fahrenheit. The laminating layer increases the weight of the glass, and thus increases the transmission loss modestly. A comparison of equivalent laminated and unlaminated glass:

Unlaminated	Laminated
1/4″ glass STC-29	2 plies 1/8″ STC-33
1/2″ glass STC-33	2 plies 1/4″ STC-36

The use of laminated glass in double-glass windows results in greater sound transmission loss because of increased mass. There is also a small damping effect, which is an advantage.

Effect of plastic panes

Plastic has the advantage of flexibility, and it does not shatter like glass. Such characteristics might suggest plastic instead of glass in the observation window in certain circumstances. The principle difference from the sound transmission loss point of view is that the mass of plastic is about half that of glass and corresponding double thicknesses of it would be required. Plastic sheets can be cold-bent to form convex windows. Light transparency of plastic is good, and optical distortion is minor. A convex form on the studio side would add a diffusing effect to incident sound and eliminate "slap-back."

Effect of slanting the glass

Old-timers in the recording field may have the idea that slanting one of the panes of a double-glass observation window is the right way of doing it. Perhaps this idea originated in attempting to control the "slap-back" reflection on the studio side. As far as transmission loss is concerned, slanting glass offers no advantage over parallel glass if the parallel glass separation is equal to the average separation of the slanting glass. Based on a great number of measurements, Quirt says "Nonparallel glazing does not appear to offer any significant benefits" (Quirt 1982).

Effect of third pane

Extensive measurements were made by Quirt on windows with a third leaf (Quirt 1983). The differences between the 3-pane windows and comparable 2-pane windows were studied, with the result that the small advantage the 3-pane had over the 2-pane was suspected of being associated with the coincidence effect. Most of the 3-pane results were very similar to those of the 2-pane. The small advantage of the 3-pane over the 2-pane window does not seem to justify the added cost and effort.

Effect of cavity absorbent

The addition of a 1-inch thickness of glass fiber around the perimeter of the interglass space was definitely shown to be advantageous (Quirt 1982). The improvement was limited to the higher frequencies, as one would expect from the characteristics of the absorbent. These tests showed the need for greater low-frequency absorbence in the cavity, but not how to achieve it.

Thermal-type glass

Two sheets of glass mounted together with an air space of the order of 1/8″ to 1/4″ forms a very effective glass for reducing heat transmission. The sound insulating properties of such glass are the same as that of a glass of the combined thickness. In other words, air spaces that small have negligible acoustical effect, and such glasses perform on the mass law alone.

Weak windows (or doors) in a strong wall

How much is a high-transmission-loss wall compromised by mounting a window (or door) in it that has a lower transmission loss? Figure 15-9 is included to make this determination a relatively easy one. Notice that the Sound Transmission Loss (STL) is used instead of STC. STC is a single-number figure used to describe a series of TL measurements made at frequency intervals (usually one-third octave) throughout the frequency range. In other words, STC is obtained by a best fit of a standard STC contour to the actual TL measurements. The graph of Fig. 15-9 applies to a single TL measurement on the TL curve. This enables the building of an accurate point-by-point graph of the new TL curve, with the weak window mounted in the strong wall. This assumes that a measured transmission-loss curve is available for both the window and the wall.

The following suggested approach uses STC figures on Fig. 15-9, which is an oversimplification stretched to the limit. Knowing that the accuracy is limited, we proceed. For an example, assume 1000 ft^2 of wall and 100 ft^2 of weaker window. Also assume that the wall area has an STC-50 rating, and the window an STC-30 rating.

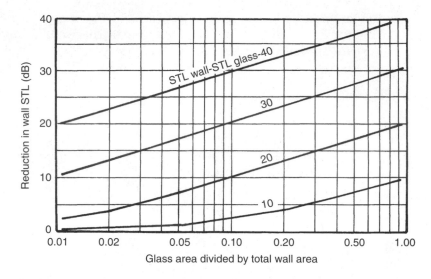

Reduction in wall STL (dB) — vertical axis
Glass area divided by total wall area — horizontal axis

STL wall-STL glass-40

30

20

10

■ **15-9** *Estimating the effect of a weak window in a strong wall.*

1. Find the ratio of the glass area to the wall area (1000/100 = 0.10), and find this ratio on the horizontal scale of the chart.
2. Find the difference between the STC of the wall and that of the glass (50 dB − 30 dB = 20 dB); and find the 20-dB STL wall-STL glass line on the graph.
3. The 0.1 vertical line hits the 20-dB line; follow it to the left scale and read 10 dB.
4. Subtract this 10 dB from the wall STC (50 − 10 = 40).

The new STC of the weakened wall is thus STC-40, which we acknowledge is a rough (but convenient) result.

Optimizing the double-glazed window

With all that has been presented above, what does it take to make a really good window? Figure 15-10 shows the measured transmission loss of an STC-55 window that has a remarkably smooth curve (Cops 1975). Surely the resonances of various kinds are well controlled to yield this smoothness. This window has dissimilar glass leaves. One leaf consists of two 3/8″ glass with a butyl layer 1/32″ thick between them. The other leaf is 5/16″. The two leaves are 12″ apart and an absorbent lines the periphery between the two leaves. The important factors are (1) large separation, (2) heavy but dissimilar panes, and (3) absorbency. Cops makes the point, however, that the absorbent produces a good improvement in sound insulation only when the mass of the panels is relatively light, which is the case for most windows. The lower curve in Fig. 15-10 is for a single 5/16″ glass pane to emphasize the great effect of adding the second, heavy leaf to accomplish the STC-55 curve. A large coincidence dip greatly reduces the STC of the single pane.

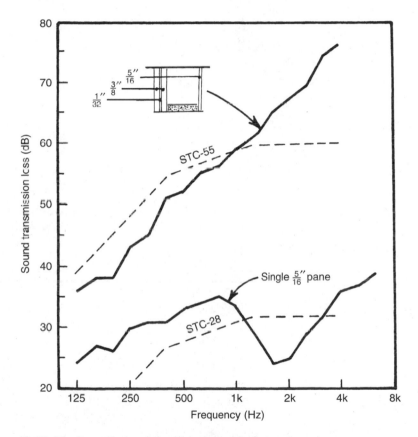

■ **15-10** *An optimized double-glass window.*

Construction of an observation window

It is not the purpose of this chapter to outline in great detail the construction of an observation window. However, the sketch of Fig. 15-11 describes the major important features:

1. A strong frame of well-seasoned 2″ × 10″ or 2″ × 12″ wood is necessary for both frame and masonry walls.

This frame must be made a part of the wall by packing glass fiber and acoustical sealant at the joints.

2. The outer stops should be screwed to the frame so that either glass can be removed for cleaning (bugs do hatch out between panes!).

3. The glass panels must be fitted with foam or rubber strips obtained from the glazier. The strip bearing the weight of the glass along the bottom should deflect about 15% under load. The strips on the other three sides can be of lighter foam, as their only function is mechanical isolation and sealing.

4. The space between the inner stops should be filled with glass fiber and covered with black cloth or perforated material offering at least 15% in openings.

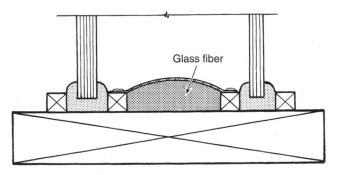

■ **15-11** *Window construction details.*

Proprietary observation windows

A well-constructed studio or other sound-sensitive space deserves the permanently tight treatment of openings that proprietary windows and doors provide. These openings are the worst place to economize, because deterioration with time is greatest in home-built doors and windows.

Figure 15-12 gives cross-sectional views of an STC-38 single-leaf window and an STC-55 double-leaf window manufactured by Overly. The transmission-loss curves for these two windows are shown in Fig. 15-13. Aside from a modest coincidence dip in the single-leaf window, the curves are very smooth.

Studio doors

The hollow-core household door of Fig. 15-14 offers only about 15 dB of attenuation over much of the spectrum, but has an STC

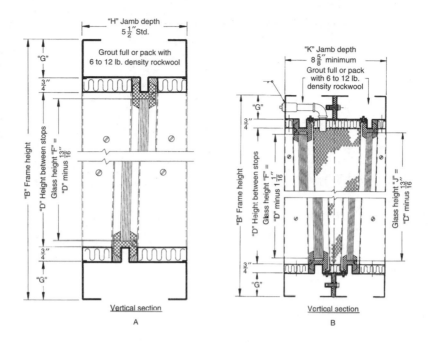

■ **15-12** *An Overly proprietary window.*

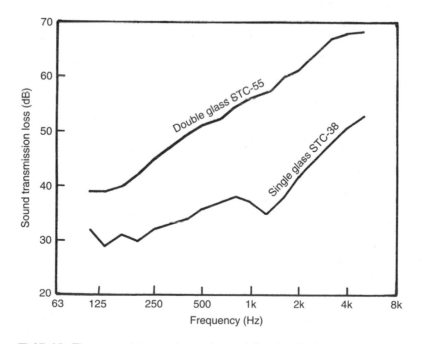

■ **15-13** *The sound transmission loss of Overly window.*

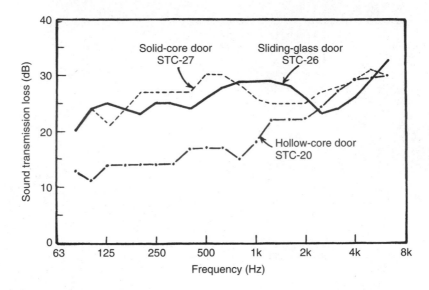

■ **15-14** *Transmission loss of a domestic door/sliding-glass door.*

of 20. Solid-core doors are somewhat better (STC-27), but still ineffective as a barrier for noise. These doors require weatherstripping, which deteriorates rapidly. The sliding-glass door, STC-26, is similar to the solid-core door. It is mentioned here because it is useful for drum cages and isolation booths. For any serious studio-type operation, household doors are too inferior to be considered and must be replaced by heavier doors. Here again, the proprietary door (as shown in Fig. 13-15) is by far the wiser way to go.

Sound-absorbing materials and structures

16

SOFT THINGS ABSORB SOUND; SO SAYS "THE PEOPLE'S Acoustics." There is a natural understanding of the deadening of sound by soft things that has grown out of personal experiences, such as burying one's head in the pillow to drown out the sound of the alarm clock. Carrying the soft idea one step further reveals the more basic concept of porosity. Porosity is readily associated with such things as rugs, carpets, and drapes, which absorb sound readily.

When sound falls on a porous surface, some of it is reflected and some absorbed. Does the absorbed part vanish? No, energy can be changed from one form to another, but it cannot be created or destroyed. Thus states the law of conservation of energy. Absorbed sound energy is changed to heat energy and it appears to vanish, because the amount of heat produced is so small it is unnoticeable. As the sound penetrates a porous material, the vibrating air particles impart movement to the tiny fibers. Such movement encounters resistance as fiber rubs on fiber, and resistance means heat.

There are many types of porous sound-absorbing materials available today. The fact that the audio band is so very wide (ten octaves!) results in a basic sound-absorption problem: All practical sound absorbers are frequency-dependent. The determining factor in absorption is the wavelength of the sound being considered. A porous absorber must be a significant fraction of a wavelength thick to be a good absorber. At 100 Hz, the wavelength of sound is 11 ft. The absorption of all porous absorbers is poor at low frequencies because of this effect.

One way to classify sound absorbers is *passive* and *active*. The porous absorber is a passive absorber. The active absorber utilizes a vibrating membrane, diaphragm, panel, or volume of air. One resonator, commonly called a Helmholtz resonator, is composed of a volume of air connected to the air of the room through a "tube" of some sort. The simplest case is a bottle in which the air in the bottle is coupled to the air of the room by the air in the neck of the bottle. If excited by blowing across the neck of the bottle, the mass of the air in the bottle and the mass of the air outside are connected by the springiness of the air. The action is similar to that of a mass on either end of a spring. In the practical Helmholtz resonator absorber, the holes in the perforated cover or the slits between the slats act as the "neck" of the resonator.

The sound absorbed by the resonator is due to the friction of the glass fiber placed in the cavity. That portion of the sound *not* absorbed is reradiated in all directions. This contributes to the diffusion of sound in the room in some small way.

Sound absorption coefficient

The *sound absorption coefficient* is a measure of the efficiency of a material in absorbing sound. If 88% of the sound energy is absorbed by a given material or structure at a given frequency, its absorption coefficient would be 0.88. One square foot of the material would give 0.88 absorption units (sabin). If it were a perfect absorber, such as the proverbial open window, each square foot would give 1.00 absorption units.

If the absorption coefficient of a material were determined, as it usually is, by laying the sample on the floor of a reverberation room, the test sound would be arriving from every angle and the resulting coefficient would be an average one. This is called the Sabine absorption coefficient. Certain esoteric questions as to whether this coefficient or the energy coefficient should be used are completely evaded by using only the Sabine coefficient in room problems. This is the one used in this book.

The absorption coefficient of a material varies with frequency, and the standard frequencies for sound absorption coefficient measurement are 125, 250, 500, 1000, 2000, and 4000 Hz. This range is far short of the audible band, but is generally sufficient for practical application.

Standard mounting terminology

Laying the sample for measurement on the reverberation room floor is intended to mimic one common way it is used in practice, cemented directly to a wall. There are many deviations from this and standard ways of referring to different mountings are needed. Table 16-1 lists the standard designations in use today.

Glass fiber availability

To give some idea of what is available, Table 16-2 lists the 700-series glass fiber products available from Owens-Corning. The Type 701 (1.5 lb/ft^3) is basically a fluffy batt useful for wall inner spaces and wherever rigidity is not required. Type 703 (3.0 lb/ft^3) is a semirigid board that cuts readily with a knife and holds its shape without support. Type 705 (6.0 lb/ft^3) is a denser and more rigid board than the 703. Glass fiber of 3 lb/ft^3 density, whether from Owens-Corning or from other suppliers, is widely used in acoustical treatment in diverse places and structures, and can be considered something of a standard in the field.

■ **Table 16-1 Common mountings for sound-absorbing materials (ASTM designation E 795-83)**

New mounting designation		Old mounting designation
A	Material directly on hard surface	#4
B	Material cemented to plasterboard	#1
C-20	Material furred out 20 mm (¾″)	#5
C-40	Material furred out 40 mm (1-½″)	#8
D-20	Material furred out 20 mm (¾″)	#2
E-405	Material spaced 405 mm (16″) from hard surface	#7

Density of absorbent

One would expect density to have an appreciable effect on the absorption coefficient. After all, a harder surface would be expected to reflect sound more readily than a softer surface. It is surprising to learn that this effect is very small over a normal range of density, as shown in Fig. 16 1. There is a small effect at frequencies above 500 Hz.

Space behind absorbent

The air space behind a glass fiber board, on the other hand, has a very great effect on the sound absorption of the board. The test board in Fig. 16-2 is one inch thick, and the density of the board is of little effect (see the previous section). The board with 0″ air space (cemented directly on the wall) will be a standard of com-

■ **Table 16-2 Owens-Corning 700 series insulation**

Technical Data

	Thickness		Width		Length		R-value
Type 701	1″	25mm	24″	609mm	48″	1219mm†	4.2
Density 1.5 pcf	1½″	38mm	24″	609mm	48″	1219mm	6.3
k-value .24	2″	51mm	24″	609mm	48″	1219mm	8.3
	2½″	64mm	24″	609mm	48″	1219mm	10.4
	3″	76mm	24″	609mm	48″	1219mm	12.5
	3½″	89mm	24″	609mm	48″	1219mm	14.6
	4″	102mm	24″	609mm	48″	1219mm	16.7
Type 711	1″	25mm	24″	609mm	48″	1219mm†	4.0
Density 1.7 pcf	1½″	38mm	24″	609mm	48″	1219mm	6.0
k-value .25	2″	51mm	24″	609mm	48″	1219mm	8.0
	2½″	64mm	24″	609mm	48″	1219mm	10.0
	3″	76mm	24″	609mm	48″	1219mm	12.0
	4″	102mm	24″	609mm	48″	1219mm	16.0
Type 703	1″	25mm	24″	609mm	48″	1219mm	4.3
Density 3.0 pcf	1½″	38mm	24″	609mm	48″	1219mm	6.5
k-value .23	2″	51mm	24″	609mm	48″	1219mm	8.7
Type 705	2½″	64mm	24″	609mm	48″	1219mm	10.9
Density 6.0 pcf	3″	76mm	24″	609mm	48″	1219mm	13.0
k-value .23	3½″	89mm	24″	609mm	48″	1219mm	15.2‡
	4″	102mm	24″	609mm	48″	1219mm	17.4‡

†Made-to-order board size. ‡Available in 703 Series Insulation only.

Made-to-order sizes are available in one-inch increments up to 48″ × 96″. Contact your local Owens-Corning sales representative for minimum order quantities.

parison. Furring a glass fiber board out from a reflecting surface greatly increases the absorbing effect. In fact, the 1″ board furred out has the absorbing efficiency of a much thicker flush-mounted board. The achievement of greater absorption can be reduced to the question of whether the furring out is cheaper than a board of greater thickness.

Thickness of absorbent

Thinking of thickness of the glass fiber board in terms of wavelength of sound is the proper approach as we examine Fig. 16-3. The wavelength of sound at 250 Hz is about 4.5 ft. The quarter wavelength is about 13 inches. The thickness of the 4″ glass fiber board is approaching the quarter-wavelength of 250-Hz sound, and the great superiority of the absorption of the 4″ board over the 1″ board is due strictly to this fact.

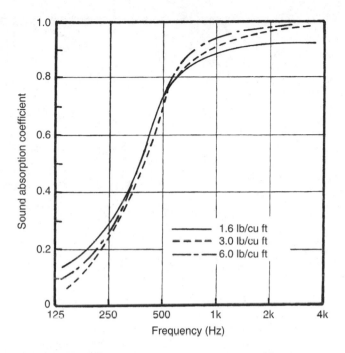

■ 16-1 *The effect of density on absorption (Owens-Corning).*

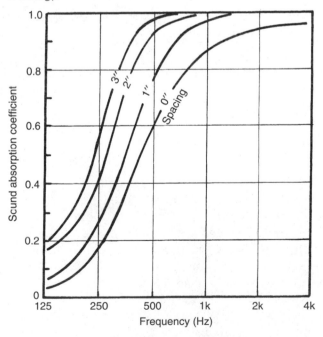

■ 16-2 *The effect of air space on glass fiber absorption (Owens-Corning).*

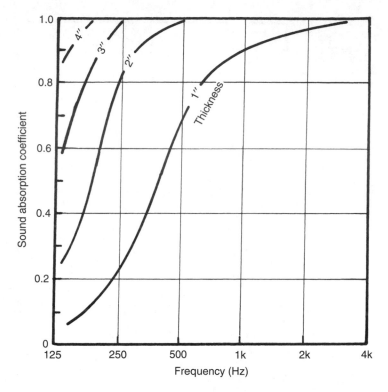

■ 16-3 *The effect of thickness of glass fiber on absorption (Owens-Corning).*

Handling 4″ glass fiberboards (and even thicker) is a problem in acoustical treatment of a studio. For one thing, space is always at a premium. Some structure will be needed to hold the 4″ or 6″ board and to protect it from damage. When absorption below 125 Hz is required, attention is usually directed toward other methods of achieving it, such as resonators.

Bass trap

There is much lore associated with the bass trap. This was largely dissipated in the early motion picture and sound recording days by the scientific approach of pioneer Michael Rettinger. He pointed out that the secret was to make the depth of the trap a quarter wavelength at the desired frequency of peak absorption. These traps were greatly exploited to achieve the "tight" bass so desired in control rooms. Traps were buried in the floor between the console and the observation window, set in the lower walls almost anywhere, and unused space in the rear or above the ceiling of the inner shell was converted to large traps. Better understanding of control room acoustics has reduced the demand for traps.

Figure 16-4 sketches the prime requisite of a bass trap, a quarter-wavelength depth. The common application of the name "bass trap" to any low-frequency absorber is confusing. The particle velocity is maximum at the surface of the trap, which means that glass fiber near that high-particle-velocity region would absorb very well. The sound pressure is zero at the surface. This low pressure means that the trap acts like a vacuum cleaner, sucking sound energy near the peak frequency into the trap from the nearby surrounding area.

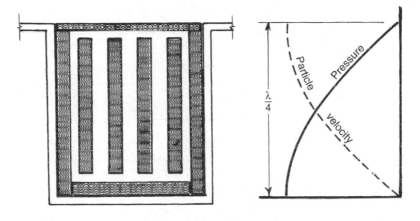

■ **16-4** *Bass trap construction.*

Circumstances usually limited the practical depth of traps to 2 or 3 ft. The 3-ft trap would peak at $1130/(3 \times 4) = 94$ Hz, while the 2-ft trap would peak at 141 Hz. These frequencies are right in the lower axial-mode range, which could well be handled by corner absorbers. Obviously, placing such a trap at the antinode of an axial mode would have little effect on that mode.

Acoustic tile

A favorite early acoustic treatment that hangs on today is to cover the walls and ceilings with $12'' \times 12''$ acoustic tile, with an average thickness of $1/2''$. These are cemented to the surface or placed in T-bar suspension systems. Even though these tiles are still available on a limited basis, information on their absorption characteristics is often not available. Figure 16-5 shows the spread of the data among eight different brands of $3/4''$ tile. Good absorption above 500 Hz is the general rule, dropping off fast below that frequency.

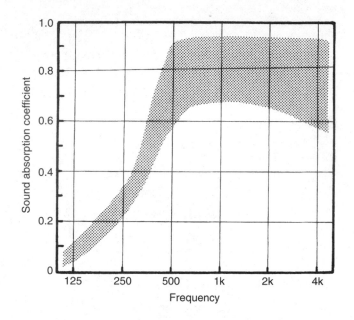

■ **16-5** *The absorption of acoustic tiles.*

Diaphragmatic absorbers

Consider a room that might become a special audio facility. The walls and ceiling are ordinary drywall panels; the floors are tongue-and-groove flooring. Sound in the room sets these surfaces to vibrating as diaphragms. The flexure of the surfaces encounters a resistance in the fibers, and heat is developed as a result of the vibration. This dissipation of heat means that sound energy is being absorbed. This sound absorption is large enough that it must be included in calculations for the room. Drywall of 1/2″ thickness on studs spaced 16″ on centers has an absorption coefficient of 0.29 at 125 Hz, and less at higher frequencies. The wall, the floor, and the ceiling are all good low-frequency absorbers, and must be considered in the calculations.

Plywood of 1/4″ thickness, spaced out from the wall is a good low-frequency absorber. The frequency of resonance of such a structure can be calculated from the expression:

$$f = \frac{170}{\sqrt{(m)(d)}} \qquad (16\text{-}1)$$

in which

f = frequency of resonance, Hz
m = surface density of the panel, lb/ft^2 of surface
d = depth of airspace, inches

For example, a 1/4″ plywood is spaced out from the wall on 2 × 4s. The surface density of 1/4″ plywood is 0.74 lb/ft^2. What is the frequency of resonance of the structure?

$$f = \frac{170}{\sqrt{(0.74)(3.75)}}$$

$$f = 102 \text{ Hz}$$

Even this minor labor of calculating the resonance frequency of this 1/4″ panel can be avoided by using the graph of Fig. 16-6, which is developed from Equation 16-1. These panel resonators can reach a relatively sharp peak of absorption coefficient of about 0.6 at their frequency of resonance. This peak can be lowered and broadened by placing some glass fiber in the space.

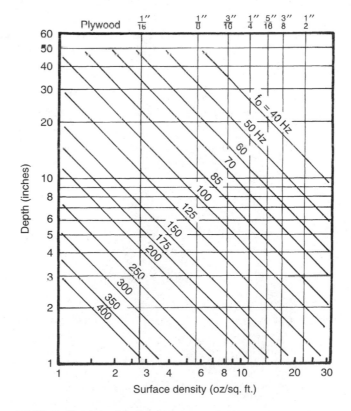

■ **16-6** *Design chart panel resonators.*

Panel absorbers

One method of making a practical resonator of a plywood panel is illustrated in Fig. 16-7. This 1/4″ plywood panel furred out on 2 × 4s (net spacing about 3-3/4″) resonates at 102 Hz according to our example above. The space would be left empty if a sharp peak of absorption is desired or flattened down with glass fiber if a flatter characteristic is desired. A 1/4″ to 1/2″ air space between the panel and the absorbent is recommended to avoid interference with the vibration of the panel. Some would suggest that the periphery be isolated from the frame with strips of gummed rubber to avoid rattles.

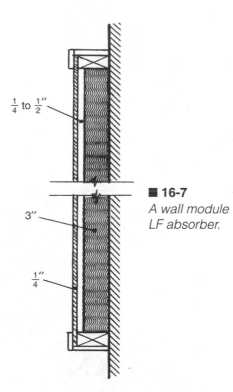

$\frac{1}{4}$ to $\frac{1}{2}$″

3″

$\frac{1}{4}$″

■ **16-7**
*A wall module
LF absorber.*

The panel resonator can be built into a corner as shown in Fig. 16-8. A very rough estimate of the frequency of resonance can be made by taking the depth of the corner as the average depth. This would usually place the resonance frequency below 100 Hz which would offer some control of axial modes. More than that, providing absorption in a corner would touch all modes of the room since that is where they all terminate.

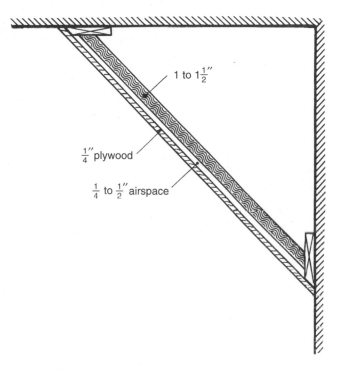

■ **16-8** *Corner modular LF absorber.*

Polycylindrical absorbers

Polycylindrical absorbers (polys) have had their day in early radio and recording studios. They have served well because their characteristics are very good. Before the day of number-theory diffusors, they offered good diffusion. They offer good low-frequency absorption, and they are dramatic in appearance. They also offer the possibility of rattles in tune to the music if not built properly. Some would say they look "old-fashioned." Time will tell if their popularity will return. In spite of the uncertainty of their future, they are mentioned here in the interest of completeness.

Anyone with experience in carpentry will be curious as to how they are constructed. Figure 16-9 offers at least a hint. The space behind the cylindrical surface is broken up with bulkheads, over which the skin is stretched. The top edge of each bulkhead has a strip of felt or foam to assure that the skin will not rattle against the bulkhead edge. The secret is in careful measurement of all elements, and accurate sawing of slots in the edge strips with a radial saw.

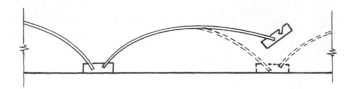

■ **16-9** *Construction details of a polycylindrical diffusor.*

Perforated panel absorbers

Perforated panel resonators are truly Helmholtz resonators, although there is little similarity in appearance between today's resonators and the metal spheres Helmholtz used in another era. Here's the idea: blowing across the neck of a bottle produces a tone. The air in a bottle is springy, and the mass of the air in the neck of the bottle reacts with the springiness to form a resonating system that oscillates at a frequency determined by the volume of the bottle and the diameter and length of the neck of the bottle.

Drilling a panel full of holes can provide the necks of many bottles. Fixing the panel over a box provides a space behind the perforated panel so that each hole (neck) has its apportioned space, completing the bottle analogy. The frequency at which the perforated cover on the box resonates can be calculated by the expression:

$$f = 200 \sqrt{\frac{p}{(d)(t)}} \qquad (16\text{-}2)$$

in which

f = frequency of resonance, Hz
p = perforation percentage,
  = [(hole area) ÷ (panel area)] 100
t = effective hole length, inches, with correction factor applied,
  = (panel thickness) + (0.8) (hole dia.)
d = depth of air space, inches

The perforation percentage can be calculated from Fig. 16-10.

Equation 16-2 is true only for circular holes. This equation is shown in graphical form in Fig. 16-11, which is specific for panels of 3/16″ thickness. Table 16-3 lists the hole diameter and spacing to yield various peak frequencies for panel thickness of 1/8″ and 1/4″ for depth of air space of 3-1/8″ and 5-1/8″. In the form of an independent module, a perforated face resonator could be constructed as shown in Fig. 16-12.

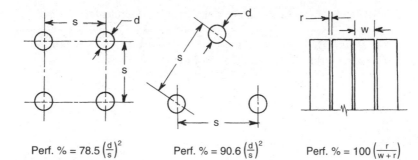

Perf. % $= 78.5 \left(\dfrac{d}{s}\right)^2$ Perf. % $= 90.6 \left(\dfrac{d}{s}\right)^2$ Perf. % $= 100 \left(\dfrac{r}{w+r}\right)$

■ **16-10** *Percent perforation computation.*

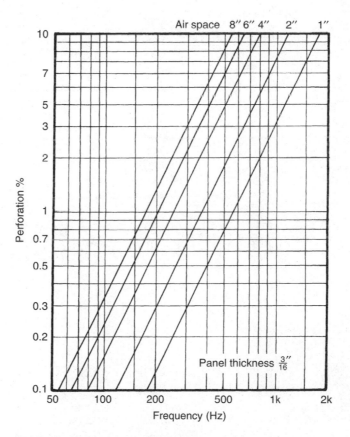

■ **16-11** *A design chart for perforated absorbers.*

■ Table 16-3 Low-frequency absorber, perforated-face type

Depth of airspace	Hole dia.	Panel thickness	% Perf.	Hole spacing	Freq. of resonance
3⅝″	⅛″	⅛″	0.25%	2.22″	110 Hz
			0.50	1.57	157
			0.75	1.28	192
			1.00	1.11	221
			1.25	0.991	248
			1.50	0.905	271
			2.00	0.783	313
			3.00	0.640	384
3⅝″	⅛″	¼″	0.25%	2.22″	89 Hz
			0.50	1.57	126
			0.75	1.28	154
			1.00	1.11	178
			1.25	0.991	199
			1.50	0.905	217
			2.00	0.783	251
			3.00	0.640	308
3⅝″	¼″	¼″	0.25%	4.43″	89 Hz
			0.50	3.13	126
			0.75	2.56	154
			1.00	2.22	178
			1.25	1.98	199
			1.50	1.81	217
			2.00	1.57	251
			3.00	1.28	308
5⅝″	⅛″	⅛″	0.25%	2.22″	89 Hz
			0.50	1.57	126
			0.75	1.28	154
			1.00	1.11	178
			1.25	0.991	199
			1.50	0.905	218
			2.00	0.783	251
			3.00	0.640	308
5⅝″	⅛″	¼″	0.25%	2.22″	74 Hz
			0.50	1.57	105
			0.75	1.28	128
			1.00	1.11	148
			1.25	0.991	165
			1.50	0.905	181
			2.00	0.783	209
			3.00	0.640	256
5⅝″	¼″	¼″	0.25%	4.43″	63 Hz
			0.50	3.13	89

Depth of airspace	Hole dia.	Panel thickness	% Perf.	Hole spacing	Freq. of resonance
			0.75	2.56	109
			1.00	2.22	126
			1.25	1.98	141
			1.50	1.81	154
			2.00	1.57	178
			3.00	1.28	218

Figures 16-13 and 16-14 are companions, the former for an air space depth of 4″ with 2″ absorbent and the latter for an air space depth of 8″ and a 4″ absorbent. Other things remaining fixed, the first thing noticed is that the smaller the perforation percentage, the lower the frequency; frequencies range from 300 Hz for the shallower unit to well below 100 Hz for the deeper unit. In the usual situation, the designer would place the peak absorption of

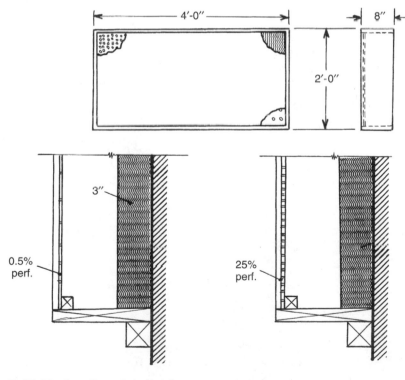

■ **16-12** *A wall module low-frequency absorber.*

the low-frequency perforated unit to compensate for the low-frequency deficiencies in other absorbers, in order to achieve flat absorbence across the audio band.

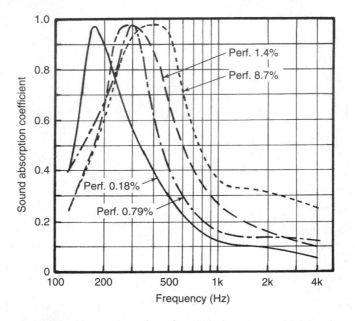

■ **16-13** *Absorption of perforated absorbers 4" in depth (Mankovsky 1971).*

Slat absorbers

In Fig. 16-10 the equivalent of perforation percentage for slat resonators is the slot area percentage. Slat resonators are full-fledged Helmholtz resonators, the hole in the neck of the bottle now being a short section of a slot with its proportional airspace below. The mass of the air in the slot reacts with the springiness of the air in the cavity. Whether perforated or slat styles are used, they are identical in results. The frequency of resonance in the slat absorber can be estimated from:

$$f = 216 \sqrt{\frac{p}{(d)(D)}}$$ (16-3)

in which

f = frequency of resonance, Hz
p = slot area percentage (see Fig. 16-10)
D = airspace depth, inches
d = thickness of slat, inches

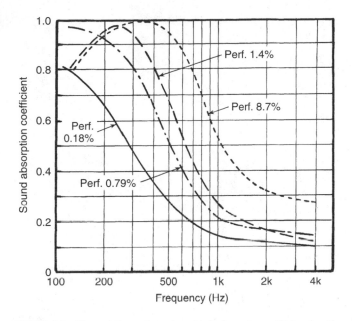

■ **16-14** *Absorption of perforated absorbers 8" in depth (Mankovsky 1971).*

The Flutter-Free diffusor strips of chapter 12 can be used as slats on a Helmholtz resonator, as shown in Fig. 12-22. The low-frequency absorption of the resonator is supplemented by high-frequency diffusion. This type of slat would eliminate any problems of troublesome specular reflections from the slat surface. The resonator could be built and positioned to provide needed low-frequency absorption and eliminate a flutter echo at the same time.

Open-cell foams

Foams are finding increasing applications as sound absorbers in sound recording and home applications. The closed-cell foams do not absorb sound, and are to be avoided.

Open-cell foams, however, are good sound absorbers, comparable to glass fiber in efficiency. Closed-cell and open-cell foams are similar in appearance, but can be identified by trying to blow air though them: air goes through the open-cell type, but not the closed-cell type.

Sonex

Sonex molded foams have been on the market for many years and are growing in popularity. In Fig. 16-15 are shown four of the

Sonex patterns: (A) SONEXpyramid, (B) SuperSONEX1, (C) SONEXone, and (D) SONEXclassic.

The absorption characteristics of SONEXpyramid is shown in Fig. 16-16. These curves can be compared directly to those of glass fiber in Fig. 16-2. It is noted that the 2″ Sonex has almost the same absorption as the 1″ glass fiber and the 4″ Sonex is close to the 2″

■ **16-15(A)** *SONEXpyramid.*

■ **16-15(B)** *SuperSONEX1.*

Sound-absorbing materials and structures

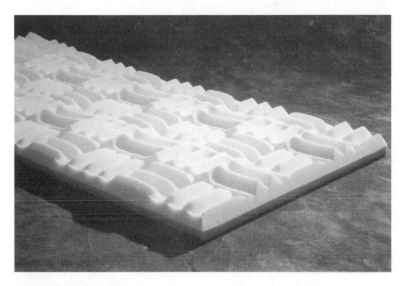

■ **16-15(C)** *SONEXone.*

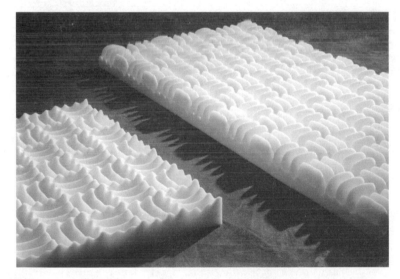

■ **16-15(D)** *SONEXclassic.*

glass fiber thickness. The Sonex thickness includes the contour crests while glass fiber is solid fibrous material, making accurate comparison difficult.

The absorption characteristics of SuperSONEX is shown in Fig. 16-17 for thickness of 2″ and 6″. Again, compared to common glass fiber of Fig. 16-2, half the thickness of Sonex will give about the same absorption as glass fiber, but the same qualifications apply: the peaks of the contour are included in Sonex measurement of thickness.

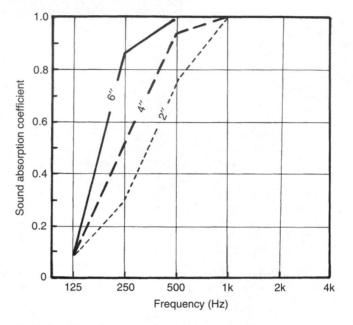

■ **16-16** *The absorption of SONEXpyramid.*

The value of Sonex is that it is a complete product that is ready to apply to the wall, while glass fiber requires framing and a cover. Some consumers appreciate the dramatic appearance of Sonex, and use it to enhance the appearance of their studio.

Are glass fibers dangerous to health?

On June 24, 1994, the U.S. Department of Health and Human Services made the announcement that fiberglass will be listed as a material "reasonably anticipated" to be a carcinogen (Hirschorn 1994). This has been a highly controversial subject because of the association with asbestos. More than 400 articles and reports have been published on the subject. Among the results were the following:

1. A 15-year epidemiological study found no excess of respiratory disease among fiberglass workers.

2. Los Alamos National Laboratory, in a two-year study, found no unusual disease patterns in laboratory animals after subjecting them to inhalation of glass fibers.

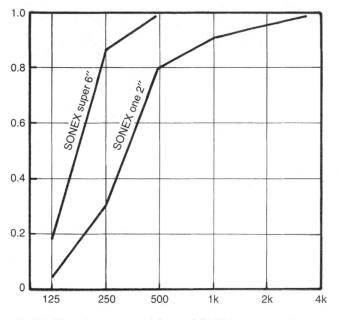

16-17 *The absorption of SuperSONEX.*

3. The World Health Organization concluded that glass fiber has not been a hazard to humans, but classified it as a "possible human carcinogen" because of studies that injected glass fibers (and many other things) artificially into animals, which produced tumors.

Despite the lack of positive evidence incriminating glass fiber, the industry is "running for cover" and making their products so that the glass fiber is contained. Owens-Corning has even established an indemnification program to protect those involved in the "specification, application, and use" of their product (Owens-Corning Commercial Building Insulation, 1995, Owens-Corning world headquarters, Fiberglas Tower, Toledo, Ohio 43659). This is simply a warning. The best approach is to wait until some hard data arrives showing this product to be harmful to humans.

Silencing the
HVAC system

IN THE ACOUSTICAL DESIGN OF STUDIOS, CONTROL ROOMS, and other sound-sensitive spaces, the primary noise (once the structure provides security from outside noises) is that which is generated and distributed by the heating, ventilating, and air-conditioning (HVAC) system. These systems generally consist of an air-moving device with some form of attached ductwork. Each system has one major noise source (e.g., the fan) along with secondary sources (e.g., grille noise) throughout the system.

The background noise in studios, listening rooms, and other sound-sensitive rooms is an ever-growing problem. Digital recording and reproducing techniques at both production and consumer levels are demanding lower and lower noise levels.

The control of HVAC noise can be expensive. The logical starting point is inserting in the air-conditioning contract a specification that a certain low-noise level must be achieved. This is a signal to the experienced contractor that there might be trouble in meeting such a specification and the contract price will be boosted accordingly. It will also be a signal to the contractor that the client means business and will be strict in demanding that the low-noise specification be met. There must be knowledgeable liaison provided by the client to check the contractor's work during progress of the job. A good engineer-type person on the client's staff designated for this liaison could and should educate himself in HVAC terminology and basic systems. An excellent way of doing this is to obtain from the American Society of Heating, Refrigerating, and Air-Conditioning Engineers a copy of their ASHRAE Handbook, and study the portions on Systems and Fundamentals. It is all there in understandable textbook form and backed by the authority of the Society.

Selection of noise criterion

There are several different ways of stating HVAC noise design goals. The single number reading of a sound level meter using the A-weighted scale is useful for noncritical systems (see chapter 11). This dBA reading discriminates against the low frequencies, more or less the way the human ear does. Obviously, this method is applicable only for undemanding situations.

The use of the NC or NCB curves (Figs. 11-8 and 11-9) in the establishment of low-noise design goals has been discussed in regard to background noise in recording studios and other critically sound-sensitive rooms. This system refers all measurements to the ultimate sensitivity of the human ear. The final question must be, "Can the background noise be heard in the room, or on recordings made in the room?"

HVAC engineers have taken this one step further in asking the question, "Does the noise sound balanced?" This introduces the quality of background noise sounds. Perhaps this approach is more applicable to an open plan office than to a recording studio, but is it really? *IF* a noise is audible in the studio or on a recording made in the studio, the quality of the noise becomes important. For instance, a pronounced rumble or excessive hiss draws attention to the noise, while a carefully balanced rumble/hiss would be less noticeable. At least this much can be agreed upon: If the background noise is audible at all, a balanced noise is far preferable to a pronounced rumble or hiss.

HVAC engineers have devised their own NC/NCB criterion curves and have called them RC (room criteria) curves. The RC curves are shown in Fig. 17-1. They have been straightened out to give a balanced audible importance to the rumbles (caused by the fan) and the hisses (caused by the diffusers). Their use is illustrated in Fig. 17-2. The fan sound spectrum peaks in the 63–250-Hz region. The diffuser sound is most important above 250 Hz. The beautiful part is that the attenuation of the two are separately adjustable, which is the key to balancing the sound. If the RC-35 contour is the design goal (as in Fig. 17-2), fan attenuation is adjusted to meet the RC-35 criterion, and the diffuser criterion can be independently adjusted to meet the RC-35 criterion. Of course, all this is done on paper in the design stage.

In the NCB criteria curves of chapter 11, the threshold of hearing is 10 dB or so below the NCB-15 contour. This means that with an

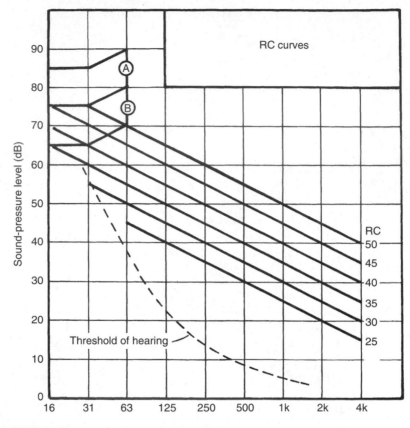

■ 17-1 *Room criterion curves.*

NCB-15 studio criterion the background noise is audible, but barely. A balanced background noise would surely be the economical approach to avoid building in too much of either fan or diffuser noise attenuation.

Fan noise

The specific sound power level of pressure blowers and centrifugal fans is shown in Fig. 17-3. In general, the pressure blowers are the noisier of the two. It is interesting to note that smaller pressure blowers are noisier that larger ones. The centrifugal fans are quieter than the pressure blowers, and the larger ones are noisier than smaller ones.

The noise of a fan results from several noise-generating mechanisms. A sirenlike tone results from interactions between the rotating and stationary members of the fan. The noise component

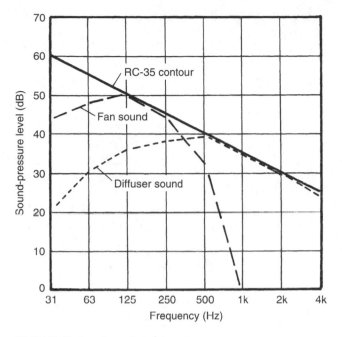

■ 17-2 *Balancing of noises.*

consists of one or several pure tones. The blade-passage frequency of fan noise can be found from the product of the revolutions per second and the number of blades. Fans with less than 15 blades produce relatively pure tones that tend to dominate the spectrum. There is more to fan noise than tones: In addition to

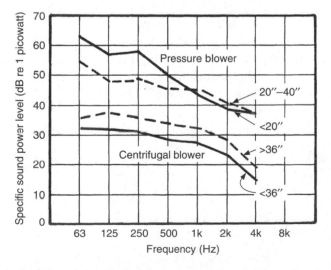

■ 17-3 *Pressure blower vs. centrifugal noise.*

this is the *random noise* generated by the vortex of air created by fan motion.

Machinery noise

The noise of the drive machinery of an HVAC system might be distributed through the structure. If the machinery is mounted on a concrete slab, it is well to isolate the local HVAC slab from the main slab. This is done during the pouring process by installing a compressed glass fiberboard between the two slabs. Locating the machinery some distance from the sound-sensitive area is highly desirable. Locating it on the roof of a frame structure is usually the least desirable, and the use of vibration mounts often ends in failure. Vibration mounts of the right design are extremely useful.

Air velocity

Noise is generated by the flow of air in ducts, and the general noise level varies approximately as the 6th power of the velocity of the air flow. If air velocity is doubled, the noise level at the outlet will increase about 16 dB. Some authorities claim that airflow noise varies as the 8th power of the velocity. In this case, doubling airflow velocity would increase the noise 20 dB.

Quantity of air delivered is a basic design parameter that determines the size of the duct. If the cross-sectional area of the duct is 1 ft^2, to deliver 500 ft^3/min requires an air velocity of 500 ft/min. Doubling the cross-sectional area of the duct to 2 ft^2, the air velocity is decreased to 250 ft/min. Budget HVAC jobs naturally incline toward higher-velocity air, smaller ducts, and higher noise levels. An air velocity of 500 ft/min is suggested as the maximum for sound-sensitive rooms.

Duct-fitting aerodynamic noise

Smooth aerodynamics of the air-flow path result in low noise. Fittings that produce gusty and swirling air flow result in higher noise generation. Aerodynamic noise is produced at elbows, dampers, branch take-offs, sound traps, and the like.

The intensity of noise generated by air flow varies as the 5th to 6th power of the velocity, and small irregularities can cause significant increases in noise.

Natural attenuation

To avoid overdesigning, the natural attenuation of a duct system must be taken into account. Duct wall vibration tends to reduce the intensity of noises generated within the system. Combined effects of duct wall vibration, energy division at branch take-offs, and reflections at elbows and duct outlets constitute a noise attenuation that must not be neglected. For example, a round duct with or without thermal insulation has a natural attenuation of about 0.03 dB/ft at 1 kHz, rising to about 0.1 dB/ft at higher frequencies.

Basic silencers

Six basic HVAC noise silencers are sketched in Fig. 17-4. The *lined duct* uses an absorptive lining to attenuate the sound. The *reactive expansion chamber* is an irregularity in the duct that reflects energy toward its source. The reverse phase of the reflection cancels some of the oncoming sound energy. The *tuned stub* is designed to resonate at the frequency of a tonal component of sound in order to cancel some of its sound energy. It, too, is a very sharp irregularity that reflects and cancels some of the undesired sound energy. The sound flows into the *sound plenum* at the input. Some of the sound energy is absorbed by the lining, and what is left escapes through the outlet in attenuated form. The *lined bend* forces the sound to turn the corner through many reflections, each being attenuated by the absorptive lining. The *diffuser* does not reduce noise, but it does slow down the high velocity air.

Lined duct

The lined duct is also called the *parallel baffle silencer*, because the absorbent is parallel to the air flow. This form of attenuator is widely used and easily implemented by the simple addition of absorbent to the duct system. The performance of this type of attenuator depends on the length of absorbing treatment, thickness of absorbent, and the acoustical characteristics of the absorbent and its perforated-metal facing material. In Fig. 17-5* the same percentage of free area is maintained in all cases. The thinner absorbent in the closer spacings provides more high-frequency attenuation than the thicker absorbent in the wider spacings. The absorbent is too thin to realize much low-frequency attenuation in any of the three cases. Similar information is presented in

* Figures 17-5 through 17-13 are patterned after figures in the ASHRAE Handbook.

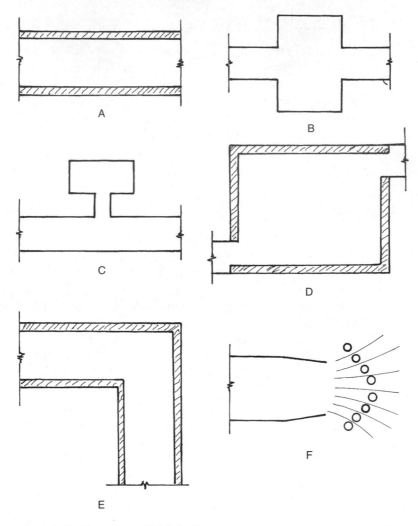

■ **17-4** *Basic types of HVAC silencers.*

Fig. 17-6 for somewhat smaller ducts lined with 1″ of absorbing material, and with the attenuation expressed in dB/ft.

Lined duct; blocked line-of-sight

Figure 17-7 shows an adaptation of the straight-lined duct type of parallel baffle. By arranging a blocked line-of-sight some of the economic advantages of thick panels are retained, as better high-frequency attenuation is achieved. Forcing the sound to go around the obstruction acts like a lined bend. These are proprietary attenuators and are available commercially.

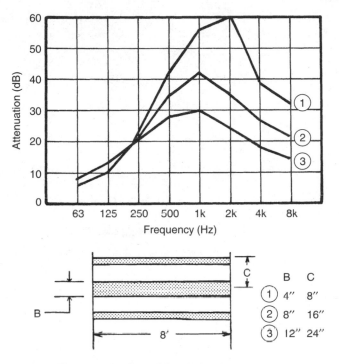

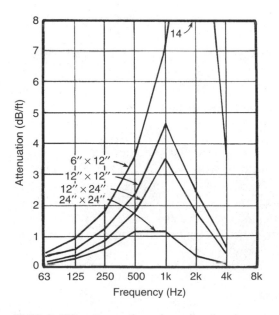

■ **17-5** *The attenuation of lined ducts.*

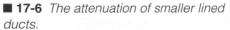

■ **17-6** *The attenuation of smaller lined ducts.*

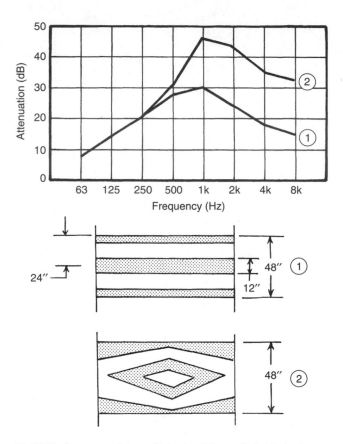

■ **17-7** *A comparison of blocked line-of-sight ducts and straight-through ducts.*

Lined duct; length effect

Figure 17-8 shows the effect of length on the attenuation performance of the lined duct arrangement of Fig. 17-5. It is interesting to note that the first four feet of treatment provides more attenuation than succeeding four-foot sections. This is explained by the fact that some of the sound energy entering the duct is in cross-mode form, not parallel form. The cross-mode energy is rapidly absorbed in the first few feet of absorbent, leaving the parallel-mode energy to continue down the duct.

Plenum chambers

Attenuating the noise energy of the machinery by lining the ducts and doing other things downstream is one approach but it might

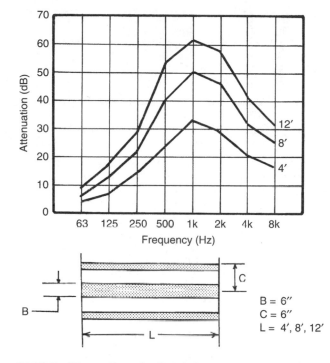

■ 17-8 *Attenuation of a lined duct.*

not be the most economical. A plenum of the form of Fig. 17-9 can be placed in the machinery room and greatly attenuate the machinery noise before it reaches the duct system. It provides attenuation as an expansion chamber in a reactive muffler and reflects energy back toward the source at both the entrance and exit irregularities. In addition, the plenum is lined with absorbing material, which absorbs both sound energy coming in and sound energy reflected from the exit irregularity. Attenuation of the plenum can be increased by increasing the ratio of the cross-sectional area of the plenum to the cross-sectional area of the entrance and exit openings. Thicker absorbent also increases the attenuation. Often a room near the machinery is converted to operate as a plenum.

Treated elbows

Reflection of sound waves from the surface is one method sound travels around a duct bend. This effect is most important at frequencies at which the wavelength is small compared to the duct dimen-

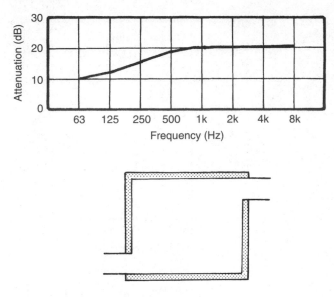

■ 17-9 *Attenuation of plenum chambers.*

sions. Reflected sound is absorbed if the bend is lined with absorbing material. The general attenuation characteristics of the lined bend are shown in Fig. 17-10. Thicker lining will improve this attenuation.

Diffuser

The sound attenuation of a typical diffuser is shown in Fig. 17-11. The effect of attenuation is achieved in the process of reducing the velocity of an air stream. This is important, because the noise of an air stream is proportional to the 8th power of the velocity. A small reduction in the velocity results in a large reduction in air stream noise. This is the only device considered so far that reduces noise better in the low frequencies than the higher frequencies.

Reactive expansion chamber

Any irregularity in the characteristics of an electrical transmission line reflects energy back toward the source. A duct carrying sound acts in the same way. The expansion chamber of Fig. 17-12 is certainly an irregularity in the acoustical characteristics of the duct. In fact, both the input and the output reflect energy back toward the source. One object is to make the expansion chamber width a quarter wavelength (at the frequency of interest) so that reflections from the output will cancel energy because of its opposite phase.

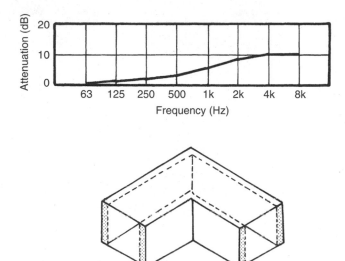

■ **17-10** *Attenuation of lined bends.*

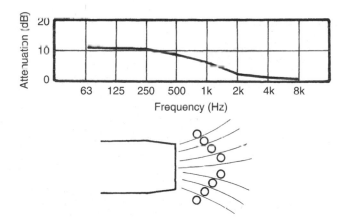

■ **17-11** *Attenuation of a diffuser.*

A chamber having the proper dimensions to cancel energy at one frequency will have no such effect on other frequencies. Therefore, it is necessary to tune the chamber to attenuate single-frequency components of the noise. This tuning will place the first maximum of attenuation at the quarter-wavelength point. As a dividend, attenuation peaks through the spectrum will also appear at frequencies of odd multiples of quarters of a wavelength. Several expansion chambers might be used in series, each tuned to differ-

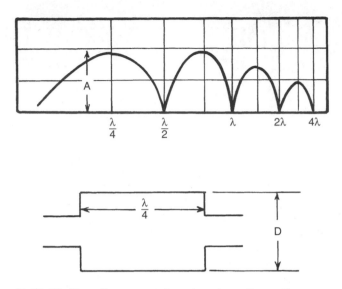

■ **17-12** *Reactive expansion chamber attenuation.*

ent noise components. No acoustical material is needed to obtain this attenuation.

Tuned stub

Another reactive attenuator relies on an acoustically tuned-stub opening into the duct as shown in Fig. 17-13. This resonator attenuates only a small band of frequencies, and has no multiples appearing down the spectrum. Even a small unit of this type can produce 40 or 50 dB of attenuation at the frequency of resonance with negligible effect at other frequencies. It offers little obstruction to air flow. Several stubs can be placed in series to attenuate other noise components.

Active noise cancellation

Why not cancel noise with a loudspeaker driven by an out-of-phase signal? This is not a new idea, but the development of signal-processing systems is encouraging its resurrection. The general plan is shown in Fig. 17-14. Noise coming down the duct is picked up by a microphone and processed, a new signal (including a reversal of phase) is generated, and that signal is used to drive a loudspeaker downstream. The sound from the loudspeaker cancels the noise...more or less. Such systems have been successfully applied to single points. In a factory, for example, the noise level at

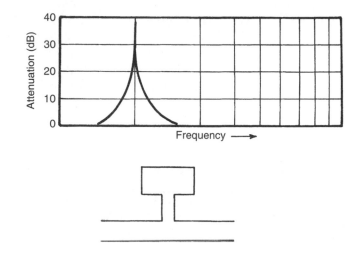

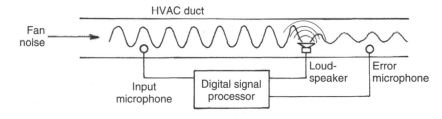

■ **17-13** *Tuned-stub attenuation.*

the telephone might be so high that conversation is difficult; a usable reduction can be obtained at the telephone by acoustical noise cancellation. Sound-canceling headphones have been quite successful. Applying the principle to HVAC systems is being done on a limited basis, with some promise for the future. Much depends on the signal-processing, and rapid strides are being made in this area.

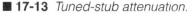

HVAC duct

Fan noise

Input microphone

Digital signal processor

Loud-speaker

Error microphone

■ **17-14** *Active noise cancellation.*

There are energy losses throughout an HVAC system even without duct lining or sound traps. This amounts to a natural attenuation which must be accounted for, lest the system be overdesigned. Only a fraction of the acoustical energy generated by the fan, the duct fittings, etc., actually reaches a given room. Branch take-offs, the vibration of duct walls, and reflection losses at bends and duct outlets all contribute their bits of attenuation. The attenuation for round ducts is about 0.03 dB/ft below 1 kHz, rising regularly to 0.01 dB/ft at high frequencies. These types of natural attenuation must be taken into consideration in the design of the system.

Without proper supervision, weird things can happen. A person unfamiliar with acoustical noise problems could very well place two outlets as in Fig. 17-15(C) without suspecting that he or she would be destroying the insulating effectiveness of an important wall between a control room and a recording studio. This is an example of the way the client's liaison with the HVAC contractor might save the day. Widely separating the outlets as in Fig. 17-15 (A) and (B) could accomplish this separation. The isolation between outlets is greater in (B) than in (A), but the attenuation of the lined duct might be sufficient.

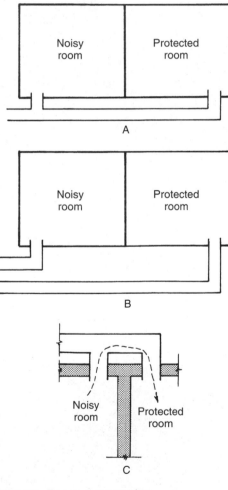

■ **17-15** *Room duct routing.*

Room acoustics and psychoacoustics

18

THE ACOUSTICAL PROPERTIES OF A ROOM ARE A DIRECT result of reflections of sound from the surfaces of the room. Signal energy radiated from a source will travel in different directions and be reflected from all six surfaces of a rectangular room. Not all the sound energy survives as sound; a small amount is converted to heat at each impact. The remainder comes together in a very complicated way to form the sound field of the room. If the source is a loudspeaker and the sound is Verdi's *Requiem Mass*, attention is all on the music, not on the mechanism of forming the intricate, detailed fluctuations of sound pressure at the ears of a listener. Such fluctuations test the limits of the human ear as to sound intensity, pitch, and timbre, as well as the limits of the human mind in understanding the physical picture of the sound field.

It would be nice if all the physicist had to do was to trace the path of each ray of sound and, with the help of sound images in the various surfaces, compute the sound pressure at the ear of the listener. That happy condition is approached in a large room, and in a small room above a certain critical frequency defined as:

$$\text{Critical frequency} = 11{,}885 \sqrt{\frac{RT60}{V}} \qquad (18\text{-}1)$$

in which $RT60$ is the reverberation time and V is the volume of the room in ft^3. For example, a studio of 3000 ft^3 with a reverberation time of 0.5 second, the critical frequency would be about 150 Hz. For a studio twice the size (same $RT60$) it would be about 100 Hz, and for a studio half the size it would be about 215 Hz. Below these frequencies, the ray concept of sound has little meaning; resonances dominate.

In rooms the size of recording studios, listening rooms, and control rooms, the low-frequency acoustics of the space is the summation of many room-resonance effects. The first part of this chapter deals

253

with the lower audible frequency range dominated by resonances. The latter part of the chapter deals with the audibility of midband reflections.

The music excites these low-frequency room resonances in a most transient and variable way as a result of the time fluctuations of the music or speech signal. These resonances of the room must be studied, because they *are* the sound field of the room for the lower part of the audible spectrum.

Resonances in tubes

Since the earliest days of music, air-filled tubes or pipes (such as organ pipes) have figured prominently in demonstrating how sound acts in confined spaces. The elementary association of pipe length with the pitch of the resultant tone is almost instinctive. Tubes can have ends that are open, closed at one end, or closed at both ends, all of which will be examined. No matter whether a solid end is on the pipe or not, that end is a reflector. It is easy to visualize a metal end as a reflector, but an open end? Yes, the open end is a reflector, too. The sound coming down the tube "sees" a very great change in impedance where the tube ends and the great outdoors begins. That great change in impedance reflects energy back toward the source. So, open or closed, the ends act as reflectors.

A pipe with a reflector on each end (a hard one or an open end) is an acoustical cavity in which standing waves can be set up and sustained. The sound traveling toward the right end of the tube is reflected from the end, and the reflected energy travels back in the opposite direction. At the same time, a reflection from the left end travels toward the right. The left-going and the right-going waves interact with each other in such a way that *at certain frequencies* a standing wave is set up in what can be called a resonance condition.

The pipe open at both ends of Fig. 18-1 will first be examined. This tube is a convenient 10 inches long. What happens is directly related to this length. The speed of sound, the frequency of the sound and the wavelength are related in this very basic and elementary statement:

$$\lambda = \frac{c}{f} \tag{18-2}$$

in which

254

c = speed of sound, 1130 ft/sec
f = frequency, Hz
λ = wavelength, ft

It is convenient to rearrange this statement into the form:

$$f = \frac{c}{\lambda} \qquad (18\text{-}3)$$

A tube that is open at both ends (Fig. 18-1) is well ventilated and open to the surrounding air. For the sound in the pipe there is no solid end reflector, just the open air. The sound pressure at the end of the tube is at maximum value. If the frequency of the sound exciting the pipe is such that the length of the pipe is 1/2 wavelength long, a standing wave is set up and the pressure pattern within the tube would be as shown in Fig. 18-1(A), with a node at the center and antinodes at each end. The node is at minimum sound pressure, and the antinode is at maximum sound pressure.

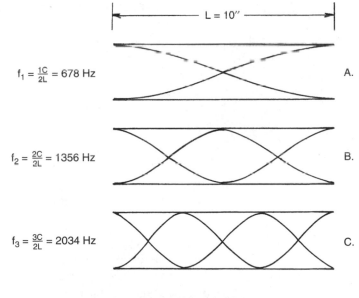

$L = 10''$

$f_1 = \frac{1c}{2L} = 678$ Hz A.

$f_2 = \frac{2c}{2L} = 1356$ Hz B.

$f_3 = \frac{3c}{2L} = 2034$ Hz C.

C = Speed of sound = 1130 ft/sec
L = Length of tube in ft = $\frac{10}{12}$ ft.

■ **18-1** *Resonances in a pipe open at both ends.*

The resonance frequency for the open-ended case of Fig. 18-1 is:

$$F = \frac{1c}{2L} = 678 \text{ Hz}$$

At twice this resonance frequency another standing wave is set up:

$$F = \frac{2c}{2L} = 1356 \text{ Hz}$$

At three times the lowest resonance frequency another standing wave is set up:

$$F = \frac{3c}{2L} = 2034 \text{ Hz}$$

At every integral multiple of the lowest resonance frequency, a standing wave is set up in tubes open at both ends.

There are significant similarities between the tube open at both ends (Fig. 18-1) and the tube closed at both ends (Fig. 18-2). In both, a standing wave can exist for every integral multiple of the lowest resonance frequency. The tube is 1/2 wavelength long at the lowest resonance frequency. A node of sound pressure *must* exist at each solid end for each standing wave for the tube closed at each end and an antinode at each end for the tube with open ends. In comparing Figs. 18-1 and 18-2, quite different pressure patterns within the tube are noted for the different standing waves, but the similarities outweigh the differences. The more basic fact is that for both ends open and both ends closed, standing waves can exist for both odd and even integral multiples of the lowest resonance frequency, and are at identical frequencies.

The case of one end open and the other end closed, shown in Fig. 18-3, is quite different from both ends open or closed. Here resonances occur only at odd integral multiples of the lowest frequency, and at the lowest frequency the tube is 1/4 wavelength long.

In all cases in Figs. 18-1 through 18-3 the pipe amounts to an enclosed space, irrespective of the end condition. These enclosed spaces resonate at some basic frequency and multiples of that basic frequency. Pipes and studios differ in many ways, but they are similar in that they are both resonant systems with standing waves forming at multiple frequencies.

Room acoustics

Turning to the acoustics of rooms, the observer is inside the pipe, so to speak. The two reflecting ends of the pipe have now become six surfaces, each of which reflects sound. As far as the simplest

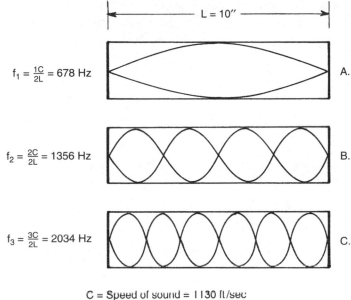

$$f_1 = \frac{1C}{2L} = 678 \text{ Hz}$$

A.

$$f_2 = \frac{2C}{2L} = 1356 \text{ Hz}$$

B.

$$f_3 = \frac{3C}{2L} = 2034 \text{ Hz}$$

C.

$L = 10''$

C = Speed of sound = 1130 ft/sec
L = Length of tube in ft = $\frac{10}{12}$ ft.

■ **18-2** *Resonances in a pipe closed at both ends.*

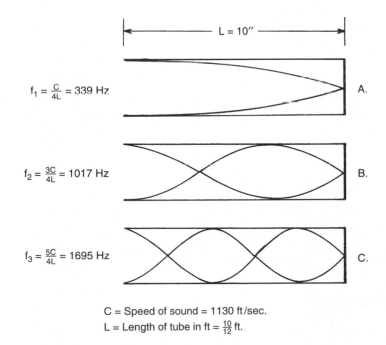

$$f_1 = \frac{C}{4L} = 339 \text{ Hz}$$

A.

$$f_2 = \frac{3C}{4L} = 1017 \text{ Hz}$$

B.

$$f_3 = \frac{5C}{4L} = 1695 \text{ Hz}$$

C.

$L = 10''$

C = Speed of sound = 1130 ft/sec.
L = Length of tube in ft = $\frac{10}{12}$ ft.

■ **18-3** *Resonances in a pipe with one end closed, one end open.*

Room acoustics

(axial) modes are concerned, the situation in the room is as though there is a horizontal pipe, a vertical pipe, and a thwartship (crosswise) pipe. Fundamental similarities exist between the pipe and the room.

Bathroom acoustics is something all have experienced. As we shall see later, bathroom acoustics will be rated very low on the scale of quality; the sound conditions in a highly reverberant bathroom can give a mediocre tenor illusions of grandeur. It will be interesting to see how room shape, the reflective/absorptive nature of the surfaces, and modal resonances cause the bathroom to sound the way it does.

An experiment in modes

A simple experiment illustrates the complexity of a sound field in a room. A loudspeaker, driven by an oscillator/amplifier combination, fills the room with the sound of a 1000-Hz tone at a comfortable level. The wavelength of this tone is 1130 ft per sec / 1000 Hz = 1.13 ft. Wild swings in loudness are heard as one moves around the room. Plugging one ear avoids complications of two-eared hearing and allows concentration on the complex changes in the sound field itself. These fluctuations in intensity result from the vector summation of many, many modal resonance components.

Resonances in a room

The pipe with both ends closed, Fig. 18-2, acts much like any opposing pair of walls in the room. There is a lowest standing-wave frequency associated with twice the length of the tube and integral multiples of that frequency.

In a room, waves can travel backward and forward between any two opposing walls. Waves can also travel at angles. At certain angles the waves return upon themselves and set up tangential and oblique standing waves. These are normal modes of vibration of the room, comparable to the normal modes of vibration in the tube.

Axial modes

There is one axial mode created by reflections from the near and far end walls of the room. This mode is actually a series of frequencies, which are the integral multiples of the lowest frequency.

Another axial mode exists between the left and right side walls of the room. This one also has a lowest frequency based on twice the distance between the two walls, and a series of integral multiple frequencies. A third set of axial modes exists between the floor and the ceiling. This one also has a lowest frequency and a series of integral multiples thereof. In summary, there are three sets of axial modes, each with a lowest frequency based on twice the distance between them, and each having a series of integral multiple frequencies at which standing waves can exist.

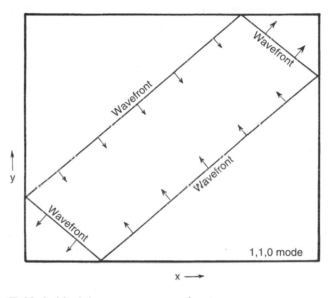

■ **18-4** *Modal resonance wavefronts.*

A word about standing wave terminology is in order here. The physicist is apt to use the standard term "stationary wave" if losses are assumed to be zero. If there are terminal losses present, the ideal situation is disturbed and the term "standing waves" applies. Surely in all our practical situations there will be losses, and the term "standing wave" is applicable.

Tangential modes

Standing waves in a room may also occur for a wave that strikes four walls and comes back to its starting place. These are tangential nodes and many of them can exist, each with its series of integral multiple frequencies (see Fig. 12-4).

Oblique modes

The oblique mode strikes all six surfaces of a room each round trip. Many oblique modes can exist and each has its series of integral multiple frequencies (see Fig. 12-4).

Figure 12-4 loses much of its rigor to oversimplification. Its chief value is graphically differentiating between *axial, tangential,* and *oblique modes.* A far more accurate way to show the forward and backward flow of sound energy in the room of Fig. 18-4 is shown in Fig. 18-5. The wavefront lines indicate planes of constant pressure extending from floor to ceiling. The forward and backward wave flow is indicated by the arrows of Fig. 18-4.

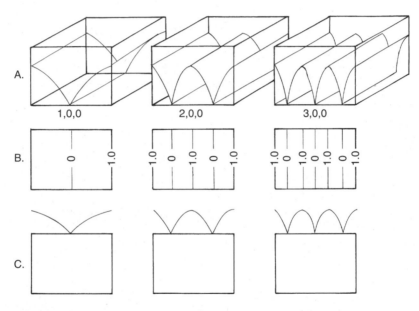

■ **18-5** *Modal sound pressure diagrams.*

Rays of sound versus waves

When the dimensions of a room are large compared to the wavelength of the sound in it (that is, for frequencies higher than the critical frequency of Eq. 18-1), the concept of sound rays has some validity. For example, at 5000 Hz the wavelength of sound is 1130/5000 = 0.23 ft. At this frequency the length of a studio could be of the order of 100 wavelengths, and the ray concept begins to make sense. However, we are not studying rays, but rather normal modes of vibration. Under these circumstances, the wave ap-

260

proach is required at low frequencies for studying the modal resonances of "small" rooms, those the size of our normal studios and listening rooms. At 20 Hz the wavelength is 1130/20 = 56.5 ft, which is far less than the critical frequency and greater than the length of the average studio.

Mode graphics

There are several methods of depicting the variations of modal sound pressures throughout a specific space. A common one is shown in Fig. 18-6 which is a plot of all the modes listed in Table 18-1.

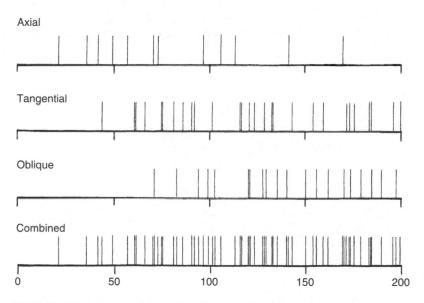

■ **18-6** *Axial, tangential, and oblique modes plotted.*

The drawings in Fig. 18-5(A) are based on vectors of sound pressure rising up from the floor to a height depending on the sound pressure at that point. In 1,0,0 there is a plane of zero pressure at the center of the room running from front to back and floor to ceiling. Maximum sound pressure (1.0 units) prevails over the entire surface of the left and right ends of the room. The floor plan of Fig. 18-5(B) is a flat-plane contour plot of sound pressure. Each one of the contour lines is the bottom edge of a plane running from floor to ceiling. For example, 0.5 pressure line shows that 50% pressure prevails everywhere on the plane from floor to ceiling.

In Fig. 18-5(C) the floor itself is left to serve other purposes while the pressure graph is moved to one side. Such a graph could appear on all four edges of the floor plan if desired. These graphs in (C) indicate the sound pressure on and above the floor area.

These same (A), (B), and (C) ideas are carried on through the 2,0,0 and the 3,0,0 modes, using exactly the same principles to convey comparable sound pressure information.

Mode calculation

The number of modal frequencies in a room may be computed from a solution of the wave equation. This was done in chapter 12, Equation 12-1. In Table 12-1 the axial, tangential, and oblique modes of a small room having the dimensions $12.46 \times 11.42 \times 7.90$ are listed and plotted in Fig. 12-6. In Table 18-1 and Fig. 18-6 the same has been done for a much larger room, one having the

■ Table 18-1 Mode calculations: room dimensions $23.3 \times 16 \times 10$ ft

Mode number	Integers p q r	Mode frequency, Hz	Axial	Tang.	Oblique
1	1,0,0	22.97	X		
2	0,1,0	35.28	X		
3	1,1,0	42.66		X	
4	2,0,0	48.60	X		
5	0,0,1	56.50	X		
6	2,1,0	60.06		X	
7	1,0,1	61.37		X	
8	0,1,1	66.61		X	
9	0,2,0	70.56	X		
10	1,1,1	70.79			X
11	3,0,0	72.80	X		
12	2,0,1	74.52		X	
13	1,2,0	74.52		X	
14	3,1,0	80.90		X	
15	2,1,1	82.46			X
16	2,2,0	85.69		X	
17	0,2,1	90.40		X	
18	3,0,1	92.15		X	
19	1,2,1	93.52			X
20	4,0,0	97.04	X		
21	3,1,1	98.67			X
22	3,2,0	101.39		X	
23	2,2,1	102.64			X

Mode number	Integers p q r	Mode frequency, Hz	Axial	Tang.	Oblique
24	0,3,0	106.00	X		
25	0,0,2	113.00	X		
26	3,2,1	116.07			X
27	2,3,0	116.61		X	
28	0,1,2	118.38		X	
29	0,3,1	120.12		X	
30	1,1,2	120.78			X
31	1,3,1	122.49			X
32	2,0,2	123.00		X	
33	2,1,2	127.97			X
34	3,3,0	128.59		X	
35	2,3,1	129.58			X
36	0,2,2	133.23		X	
37	3,0,2	134.42		X	
38	1,2,2	135.36			X
39	0,4,0	141.25	X		
40	2,2,2	141.81			X
41	4,3,0	143.71		X	
42	3,2,2	151.82			X
43	0,3,2	154.94		X	
44	1,3,2	156.78			X
45	3,4,0	158.90		X	
46	2,3,2	162.38			X
47	0,0,3	169.50	X		
48	1,0,3	171.19		X	
49	3,3,2	171.19			X
50	0,1,3	173.13		X	
51	1,1,3	174.79			X
52	2,0,3	176.33		X	
53	2,1,3	179.83			X
54	0,2,3	183.60		X	
55	3,0,3	184.47		X	
56	1,2,3	185.16			X
57	2,2,3	169.93			X
58	4,0,3	195.31		X	
59	3,2,3	197.51			X
60	0,3,3	199.92		X	
61	2,3,3	205.74			X
62	0,4,3	220.64		X	
63	0,0,4	226.00	X		
64	3,0,4	237.43		X	
65	0,3,4	249.63		X	

dimensions of $23.3 \times 15.0 \times 10$ ft. It is easy to compare the two because both are plotted to the same frequency scale. Note especially the closer spacing of the modes in the larger room. These modes constitute the complete low-frequency sound fields of these two rooms.

The integers p, q, and r of Eq. 12-1 are the only variables once the length L, the width W, and the height H are established for a given room. These integers not only determine the frequency of each mode, they provide a system of identification for every mode. If $p = 1$, $q = 0$, and $r = 0$, this is the shorthand for the 1,0,0 mode.

Modal width and spacing

Each mode in Fig. 18-6 is represented by a narrow line. Actually, as discussed in chapter 12, each mode is a resonance curve with a small, but finite, width. The bandwidth of a mode is the width of the resonance curve measured at points 3 dB down from the peak. This bandwidth is determined by the absorption of the room. For a reverberation time of 0.5 second (a reasonable time for an average studio), the bandwidth of each mode would be about 4.4 Hz.

With 60 modes under 200 Hz in the example of Fig. 18-6, the average spacing is 3.3 Hz. If the bandwidth of each mode is 4.4 Hz, it would appear that the skirts of the modal resonances overlap in a good percentage of the cases. Is this good or bad? There are comments to be made on both sides.

All modes have not been created equal. "Thus, for a given pressure amplitude an axial wave has four times the energy of an oblique wave" (Morse 1944). This means that the energy level of the axial mode is 6 dB higher than the oblique mode (10 log 4 = 6 dB). Elsewhere he states that the axial mode has an energy level 3 dB higher than the tangential mode. In looking at the combined modes in Fig. 18-6 we must realize that they are the acoustics of the room for that part of the spectrum below 200 Hz. In spite of the equal heights of the lines in this figure, tangential modes are 3 dB down in energy from the axial modes and the oblique modes are down 6 dB. This means that the relative contribution of a mode to the superimposed total curve depends upon what kind of a mode is being discussed. The energy levels of the three types of modes differ. Because the axial modes are the more potent, a casual appraisal of the axial modes alone might give a good estimate of the performance of a space under consideration.

Modal resonances separated too much from their neighbors can create problems. The response curve of a space can be treated as a vectorial superposition of all the individual resonance curves. If these resonances are spaced greatly, the room response curve will be irregular. Separation is not the only problem; damping, excitation, and phase of individual modes also affect their contribution to the overall superimposed curve in any given situation.

Modal resonances that are too close together can also create problems. Let us assume that two adjoining resonances are excited by a signal. When the signal changes in spectrum and/or amplitude, the excited modes are left "without visible means of support" to decay on their own. A mode excited by a pure tone will, when the excitation is removed, decay smoothly. Under the same circumstances, two modes close together beat with each other during the decay, resulting in an irregular decay curve. Irregular decays can be audible.

If the exciting signal is a human voice, certain modal problems will cause coloration (distortion) of the voice signal. Years ago this particular effect was studied very carefully by engineers of the British Broadcasting Corporation because of the detrimental effect on voice signal quality (Gilford 1972). A special amplifier was devised by which a narrow sliver of the spectrum could be amplified to any degree required. A voice signal that was noticeably colored (by the acoustics of the space) was passed through this amplifier. This peak of amplification was applied to the voice signal and adjusted until the coloration was clearly identified, and its frequency noted. A modal resonance irregularity was invariably identified with the voice coloration. He concluded that, for a coloration to be audible, an axial mode must be separated from its neighbors by about 20 Hz or more. He found that colorations appear chiefly in the region 75–200 Hz. Colorations are rare below 80 Hz because of the low energy content of voice signals in that region. Colorations disappear above 300 Hz because of the increased number of all modes at higher frequencies.

Gilford also discusses the work of Kuhl, as reported in a German-language journal. Kuhl's experiments identified the same colorations with reflections from untreated walls, observation windows, or corner reflections. Such early reflections are attenuated but little, and form a comb-filter spectrum with a definite pitch that colors the sound. If Kuhl is right, Gilford is wrong. This is very interesting as we study Toole's work on sound reflections later in this chapter.

Sound reflections and psychoacoustics

The importance of reflected sound in music halls has been known for many years. The contribution to the enjoyment of the music of lateral reflections is recognized to the extent that designers and builders of such halls must take special care that there be a proper supply of such reflections. There has been considerable research on the perception of reflected sound. The common physical arrangement for such research is similar to the usual hi-fi stereo listening arrangement. One loudspeaker, carrying the direct sound, is pointed at the observer. To one side, another loudspeaker emits the simulated delayed reflection. Many observations with "reflections" of various amplitudes and time delay provide the basic data from which important conclusions can be drawn.

The data of curve A (Fig. 18-7) was obtained in a similar way by Olive and Toole (1989). Using speech as the signal, the observer heard the direct sound with a "reflection" of variable amplitude and delay superimposed. The observer was instructed to tell exactly what was heard. The observer hears no trace of a reflected sound until its amplitude and delay reached curve A. This curve is called the absolute threshold of reflection audibility. In the shaded area below curve A, no reflections are audible.

Curve C defines the conditions under which the reflections are perceived as discrete echoes. This curve is a composite from data of other researchers (Lochner 1958 and Meyer 1952). Figure 18-7

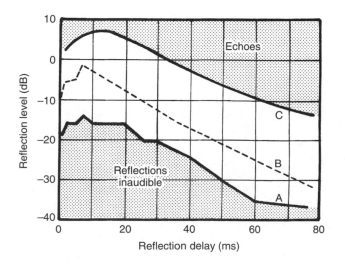

■ **18-7** *Perception of reflections.*

depicts the whole story of the effects of lateral reflections on speech signals. Below the threshold curve A, reflections are inaudible. Above curve C, reflections have deteriorated to useless, discrete echoes. The space between curves A and C is an extremely valuable and useful area that defines how lateral reflections can affect the sound in practical listening situations.

When the amplitude of reflections was about 10 dB above the threshold curve A, the observers noted that something different was happening to the direct signal. In addition to spaciousness, shifting and spreading of the front auditory image were noted, and this point has been designated as curve B. Spaciousness was sensed both below and above curve B.

It is remarkable that even though these tests were run in an anechoic room, that room was given a sense of spaciousness, of being much larger and of a different character. The observer is not aware of the reflection as a separate event; the sound of the front (direct) loudspeaker is just given this spacious character by the presence of the reflected component.

Spaciousness is a characteristic highly prized in music halls, which today are built to supply the lateral reflections necessary to create the spacious impression. Now that such reflections have been well defined (as in Fig. 18-7), the possibility arises of applying this

■ **Table 18-2 Reflection computations of Fig. 18-9**

Reflection I.D. number	Reflection path length, ft.	Reflection path length minus direct path length	Reflection level, dB	Reflection delay, ms
1. (floor)	8.1	1.0	−1.1	0.9
2. (front wall)	10.2	3.1	−3.1	2.7
3. (ceiling)	16.4	9.3	−7.3	8.2
4. (near side wall)	13.4	6.3	−5.2	5.6
5. (far side wall)	20.3	13.2	−9.1	11.7
6. (rear wall #1)	36.4	29.3	−14.2	25.9
7. (rear wall #2)	46.8	39.7	−16.4	35.1

(Direct = 7.1 ft)

$$\text{Reflection level} = 20 \log \frac{\text{direct path}}{\text{reflected path}}$$

$$\text{Reflection delay} = \frac{(\text{reflected path}) - (\text{direct path})}{1130}$$

(Perfect reflection and inverse square propagation assumed)

knowledge to a listening room. By carefully adjusting the amplitude and delay of early reflections, a spacious sense can be given to the loudspeaker sound, the room can be made to seem larger, and the stereo image can be optimized in regard to shifting and spreading of the image.

Calculating reflections

In Fig. 18-8 the knowledge of the behavior of early reflections from Fig. 18-7 is studied in a practical example. The reflections are identified by numbers, which are referred to Table 18-2. Measured direct and reflected paths are entered into the appropriate

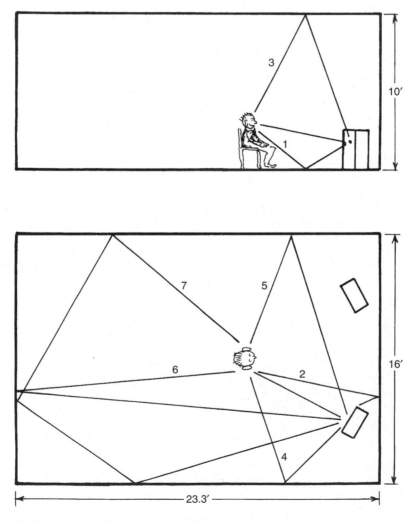

■ **18-8** *A reflection study in a listening room.*

columns of Table 18-2. Through simple computations based on perfect reflections and inverse square propagation, the magnitude and delay of each reflection are estimated. These are plotted in Fig. 18-9, and also in Fig. 18-7. It will be found that in Fig. 18-7 they are all in the area between curves A and C, some being above curve B and some below. If this were a listening position in a real room, listening tests should be conducted with these reflections specifically in mind.

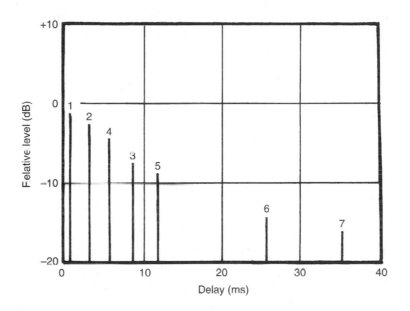

■ **18-9** *Reflections in a listening room.*

All these early reflections are adjustable in regard to amplitude, although the delay is fixed by the geometry. A given reflection can be reduced in amplitude by applying a minimum area of absorbing material at the point of impact. Squares of absorbent on the floor, the side and front walls, and the ceiling can greatly reduce or essentially eliminate the early reflections. The reader can use his or her ingenuity to adjust early reflections to achieve the sense of spaciousness and the stereo image qualities desired. Absorbers of different absorbing efficiency can adjust reflection amplitude.

With respect to early reflections, we then have a choice. Early reflections can be eliminated (as in the listening positions of chapters 1 through 10), or, for the research-minded, they can be adjusted to provide the desired stereo image and/or degree of spaciousness.

Appendix:
Absorption coefficients

Material	125 Hz	250 Hz	500 Hz	1 kHz	2kHz	4kHz	Ref.
Building Mtls.							
Concrete blk. coarse	0.36	0.44	0.31	0.29	0.39	0.25	1
Concrete blk. painted	0.10	0.05	0.06	0.07	0.09	0.08	1
Glass, large heavy panes	0.18	0.06	0.04	0.03	0.02	0.02	1
Glass, window	0.35	0.25	0.18	0.12	0.07	0.04	1
Plaster, gyp. smooth, on brick	0.013	0.015	0.02	0.03	0.04	0.05	1
Paster, gyp. smooth, on lath	0.14	0.10	0.06	0.05	0.04	0.03	1
Gypsum board 1/2″ on 2 x 4s 16″ o.c.	0.29	0.10	0.05	0.04	0.07	0.09	1
Carpet, heavy on concrete	0.02	0.06	0.14	0.37	0.60	0.65	1
Carpet, heavy on 40 oz hairfelt	0.08	0.24	0.57	0.69	0.71	0.73	1
Carpet, heavy latex backing on foam or 40 oz hairfelt	0.08	0.27	0.39	0.34	0.48	0.63	1

Material	125 Hz	250 Hz	500 Hz	1 kHz	2kHz	4kHz	Ref.
Carpet, indoor-outdoor	0.01	0.05	0.10	0.20	0.45	0.65	2
Acoustical tile, 1/2" average	0.07	0.21	0.66	0.75	0.62	0.49	
Acoustical tile, 3/4" average	0.09	0.28	0.78	0.84	0.73	0.64	
Concrete floor	0.01	0.01	0.015	0.02	0.02	0.02	1
Floor: linoleum, asphalt tile, or cork on concrete	0.02	0.03	0.03	0.03	0.03	0.02	1
Floor: wood	0.15	0.11	0.10	0.07	0.06	0.07	1
Plywood panel, 3/8" thick	0.28	0.22	0.17	0.09	0.10	0.11	1
Polycylindrical Chord 45" height 15" empty	0.41	0.40	0.33	0.25	0.20	0.22	3
Chord 35" height 12" empty	0.37	0.35	0.32	0.28	0.22	0.22	3
Chord 28" height 10" empty	0.32	0.35	0.3	0.25	0.20	0.23	3
Chord 20" height 8" empty	0.25	0.30	0.33	0.22	0.20	0.21	3
Chord 20" height 8" filled	0.30	0.42	0.35	0.23	0.19	0.2	3
Perforated panel 5/32" thick 4" depth 2" glass fiber							

Material	125 Hz	250 Hz	500 Hz	1 kHz	2kHz	4kHz	Ref.
Perf. 0.18%	0.40	0.70	0.30	0.12	0.10	0.05	3
Perf. 0.79%	0.40	0.84	0.40	0.16	0.14	0.12	3
Perf. 1.4%	0.25	0.96	0.66	0.26	0.16	0.10	3
Perf. 8.7%	0.27	0.84	0.96	0.36	0.32	0.26	3
5/32″ thick 8″ depth 4″ glass fiber							
Perf. 0.18%	0.80	0.58	0.27	0.14	0.12	0.10	3
Perf. 0.79%	0.98	0.88	0.52	0.21	0.16	0.14	3
Perf. 1.4%	0.78	0.98	0.68	0.27	0.16	0.12	3
Perf. 8.7%	0.78	0.98	0.95	0.53	0.32	0.27	3
7″ airspace 1″ min. fiber of 16 lb/cu ft density, 1/4″ cover							
Wideband 25% perf.	0.67	1.09	0.98	0.93	0.98	0.96	3
Midpeak 5% perf.	0.60	0.98	0.82	0.90	0.49	0.30	3
Lowpeak 0.5% perf.	0.74	0.53	0.40	0.30	0.14	0.16	3
2″ air space filled w. mineral fiber 9–10 lb/ cu ft density							
Perf. 0.5%	0.48	0.78	0.60	0.38	0.32	0.16	3

Key to References
(See Reference section starting on page 286 for complete citations):

1. Hedeen (1980)
2. Seikman (1969)
3. Mankovsky (1971)

Glossary

A-weighting A frequency-response adjustment of a sound-level meter that makes its reading conform, roughly, to human response.

Abffusor A proprietary panel offering both absorption and diffusion of sound.

absorption coefficient The fraction of sound energy absorbed at a given surface. It has a value between 0 and 1 and varies with frequency and angle of incidence.

absorption In acoustics, the changing of sound energy to heat.

acoustics The science that deals with the production, control, transmission, reception, and effects of sound.

active sound absorber A resonant sound absorbing structure.

AES Audio Engineering Society

ambience The distinctive acoustical characteristics of a given space.

amplitude distortion Distortion of the waveshape of the signal.

amplitude The instantaneous magnitude of an oscillating quantity such as sound pressure. The peak value is the maximum value.

analog A signal whose frequency and level vary continuously in direct relationship to the original electrical or acoustical signal.

anechoic Without echo.

anechoic chamber A room designed to suppress internal sound reflections. Used for acoustical measurements.

arrival gap The time between the arrival of the direct signal and reflections. See also *initial time-delay gap*.

articulation A quantitative measure of the intelligibility of speech; the percentage of speech items correctly perceived and recorded.

artificial reverberation Reverberation generated by electrical or acoustical means to simulate that of concert halls, etc. It is added to a signal to make it sound more lifelike.

ASA Acoustical Society of America.

ASHRAE American Society of Heating, Refrigerating, and Air-Conditioning Engineers.

attack The beginning of a sound; the initial transient of a musical note.

attenuate To reduce the level of an electrical or acoustical signal.

attenuator A device, usually a variable resistance, used to control the level of an electrical signal.

audio frequency An acoustical or electrical signal of a frequency that falls within the audible range of the human ear, usually taken as 20 Hz to 20 kHz.

audio spectrum See *audio frequency*.

auditory area The sensory area lying between the threshold of hearing and the threshold of feeling or pain.

auditory system The human hearing system made up of the external ear, the middle ear, the inner ear, the nerve pathways and the brain.

aural Having to do with the auditory mechanism.

axial mode The room resonances associated with the three axes of the rectangular room.

baffle A movable barrier used in the recording studio to improve separation of signals from different sources. The surface or board upon which a loudspeaker is mounted.

bandpass filter A filter that attenuates signals both below and above the desired passband.

bandwidth The frequency range passed by a given device or structure. Commonly measured as the width between minus 3 dB points.

basilar membrane A membrane inside the cochlea that vibrates in response to sound, exciting the hair cells.

bass boost An increase in level of the lower range of frequencies, usually achieved by electrical circuits.

bass trap A structure designed to absorb low-frequency sound energy.

bass The lower range of audible frequencies.

beats Periodic fluctuations resulting from superimposing signals of slightly different frequency.

binaural Listening with two ears.

boomy A colloquial expression for excessive bass response.

274

byte A term used in digital systems. One byte is equal to eight bits of data. A bit is the elemental "low" or "high" state of a binary system.

capacitor An electrical component that passes alternating currents but blocks direct currents. Also called a *condenser*, it is capable of storing electrical energy.

clipping An electrical signal clipped by electronic circuits or by overloading an electronic device. It is a distortion of the signal.

cochlea The portion of the inner ear that changes mechanical vibrations to electrical signals. It is the frequency-analyzing portion of the auditory system.

coincidence effect Sound energy falling on a wall having a frequency coincident with the natural period of the wall sustains the wall vibration. This results in a decrease in the transmission loss for that wall near that frequency.

coloration The distortion of a signal detectable by the ear

comb filter A distortion produced by combining an electrical or acoustical signal with a delayed replica of itself. The result is constructive and destructive interference that introduces peaks and nulls into the frequency response. When plotted to a linear frequency scale, the response resembles a comb, hence the name.

compression Reducing the dynamic range of a signal by electrical circuits that reduce the level of loud passages.

condenser See *capacitor*.

correlogram A graph showing the correlation of one signal with another.

cortex See *auditory cortex*.

critical band In human hearing, only those frequency components within a narrow band, called the critical band, will mask a given tone. Critical bandwidth varies with frequency, but is usually between 1/6 and 1/3 octave wide.

crossover frequency In a loudspeaker with multiple radiators, the crossover frequency is the -3 dB point of the network dividing the signal energy.

crosstalk The signal of one channel, track, or circuit interfering with another.

cycles per second The frequency of an electrical signal or sound wave. Measured in Hertz (Hz).

dB See *decibel*.

275

dB(A) A sound level meter reading with an A-weighting network simulating the human ear at a loudness level of 40 phons.

dB(B) A sound level meter reading with a B-weighting network simulating the human-ear response to a loudness level of 70 phons.

dB(C) A sound level meter reading with no weighting network in the circuit, i.e., flat. The reference level is 20 micropascals.

decade Ten times any quantity or frequency range. The range of the human ear is about three decades.

decay rate A measure of the decay of acoustical signals, expressed as a slope in dB/second.

decibel The human ear responds logarithmically and it is convenient to deal in logarithmic units in audio systems. The bel is the logarithm of the ratio of two powers, and the decibel is one tenth of a bel.

delay line A digital, analog, or mechanical device employed to delay one signal with respect to another.

diaphragm Any surface that vibrates in response to sound or is vibrated to emit sound, such as in microphones and loudspeakers. Also applied to wall and floor surfaces vibrating in response to sound.

dielectric An insulating material. The material between the plates of a capacitor.

diffraction The distortion of a wavefront caused by the presence of an obstacle in the sound field.

diffusion The process of scattering of sound.

diffusion coefficient The ratio of scattered intensity at ±45 degrees to the specular intensity.

diffusor A proprietary device for the diffusion of sound through reflection-phase-grating means.

digital A numerical presentation of an analog signal. Pertaining to the application of digital techniques to common tasks.

distance double In pure spherical divergence of sound from a point source in free space, the sound pressure level decreases 6 dB for each doubling of the distance. This condition is rarely encountered in practice, but it is a handy rule to remember in estimating sound changes with distance.

distortion Any change in waveform or harmonic content of an original signal as it passes through a device. The result of nonlinearity within the device.

distortion, harmonic The harmonic content of a signal is changed by passing it through a nonlinear device.

dynamic range All audio systems are limited by inherent noise at low levels and by overload distortion at high levels. The usable range between the two extremes is the dynamic range of the system. Expressed in dB.

dyne The force that will accelerate a one-gram mass at the rate of 1 cm/sec. The old standard reference level for sound pressure was 0.0002 dyne/cm^2. The same level today is expressed as 20 micropascals.

ear canal The external auditory meatus; the canal between the pinna and the eardrum.

eardrum The tympanic membrane located at the end of the ear canal that is attached to the ossicles of the middle ear.

early sound Direct and reflected components that arrive at the ear from a source during the first 50 ms or so. Such components are replicas of the original sound and arrive at different times producing comb-filter distortion. Also the basis of desirable effects such as spaciousness and sharpening of the stereo image.

echo A delayed sound that is perceived by the ear as a discrete sound image.

echograms A record of the very early reverberatory decay of sound in a room.

EFC Energy-frequency curve.

EFTC Energy-frequency-time curve.

ensemble Musicians must hear each other to function properly; in other words, ensemble must prevail. Diffusing elements surrounding the stage area contribute to ensemble.

equalization The process of adjusting the frequency response of a device or system to achieve a flat or other desired response.

equalizer A device for adjusting the frequency response of a device or system.

equal loudness contour A contour representing a constant loudness for all audible frequencies. The contour having a sound pressure level of 40 dB at 1,000 Hz is arbitrarily defined as the 40-phon contour.

ETC Energy-time curve.

Eustachian tube The tube running from the middle ear into the pharynx; it equalizes middle-ear atmospheric pressure.

external meatus The ear canal terminated by the eardrum.

feedback, acoustic Unwanted interaction between the output and the input of an acoustical system, e.g., between the loudspeaker and the microphone of a system.

FFT Fast Fourier Transform. An iterative program that computes the Fourier Transform in a relatively short time.

fidelity As applied to sound quality, the faithfulness to the original.

filter, bandpass A filter that passes all energy between a low-frequency cutoff point and a high-frequency cutoff point.

filter, high pass A filter that passes all energy above a cutoff frequency.

filter, low pass A filter that passes all energy below a cutoff frequency.

flanking sound Sound traveling by circuitous paths which reduces the effectiveness of a barrier.

floating floor A massive floor that is resiliently connected to the structure for the purpose of increasing transmission loss.

flutter A repetitive echo set up by parallel reflecting surfaces.

Fourier analysis Application of the Fourier transform to a signal to determine its spectrum.

fractal Numerous natural phenomena exhibit a macroscopic property or shape which is repeated microscopically at progressively smaller scales. These possess the property of self-similarity. Fractals applied to quadratic-residue diffusors result in extended range, similar to two- or three-way loudspeaker systems.

frequency The measure of the rapidity of alternations of a periodic signal, expressed in cycles per second or Hz.

frequency response The changes of the sensitivity of a circuit or device with frequency.

FTC Frequency-time curve.

fundamental The basic pitch of a musical note.

fusion zone All reflections arriving at the observer's ears within 20–40 ms of the direct sound are integrated, or fused together, with a resulting apparent increase in level and a pleasant change of character. This is the *Haas effect*.

gain The increase in power level of a signal produced by an amplifier.

graphic-level recorder A device for recording of signal level in dB vs. time on a tape. The level of dB vs. angle can be recorded also for directivity patterns.

grating, diffraction An optical grating consists of minute parallel lines used to break light down into its component colors.

grating, reflection phase An acoustical diffraction grating to diffuse sound.

Haas Effect See fusion effect. Also called the *precedence effect*. Delayed sounds are integrated by the auditory apparatus if they fall on the ear within 20–40 ms of the direct sound. The level of the delayed components contributes to the apparent level of the sound. It is accompanied by a pleasant change in character.

hair cell The sensory elements of the cochlea that transduce the mechanical vibrations of the basilar membrane to nerve impulses that are sent to the brain.

harmonic distortion See *distortion, harmonic*.

harmonics Integral multiples of the fundamental frequency. The first harmonic is the fundamental, and the second is twice the freqency of the fundamental, etc.

hearing loss The loss of sensitivity of the auditory system measured in dB below a standard level. Some hearing loss is age-related; some is related to exposure to high-level sound.

Helmholtz resonator A reactive, tuned sound absorber. A bottle is such a resonator. Made by placing a perforated cover or slats over a cavity.

henry The unit of inductance.

Hertz The unit of frequency, abbreviated Hz. Cycles per second.

high-pass filter See *filter, high pass*.

IEEE Institute of Electrical and Electronic Engineers

impedance matching Maximum power is transferred from one circuit to another when the output impedance of the one is matched to the input impedance of the other. Maximum power transfer may be less important in many electronic circuits than low noise or voltage gain.

impulse A very short, transient electric or acoustic signal.

inductance An electrical characteristic of circuits, especially of coils, that introduces inertial lag because of the presence of a magnetic field. Measured in henrys.

initial time-delay gap The time gap between the arrival of the direct sound and the first sound reflected from the surfaces of the room. See *arrival gap*.

in phase Two periodic waves reaching peaks and going through zero at the same time are said to be "in phase."

insulation As referred to sound barriers, insulation is applied to the sound transmission loss of a particular wall, etc.

intensity Acoustic intensity is sound energy flux per unit area. The average rate of sound energy transmitted through a unit area normal to the direction of sound transmission.

interference The combining of two or more signals results in an interaction called interference. This can be constructive or destructive. Another use of the term refers to undesired signals.

intermodulation distortion Distortion produced by the interaction of two or more signals. The distortion components are not harmonically related to the original signals.

inverse square law See *spherical divergence*.

isolation Refers to the isolation of an entire studio or room from outside noise.

ITD Initial time-delay gap.

JAES Journal of the Audio Engineering Society.

JASA Journal of the Acoustical Society of America.

kHz 1,000 Hz.

law of the first wavefront The first wavefront falling on the ear determines the perceived direction of the sound.

level A sound pressure level in dB means that it is calculated with respect to the standard reference level of 20 micropascals. The word "level" associates that figure with the appropriate standard reference level.

linear A device or circuit with a linear characteristic means that a signal passing through is not distorted.

live end/dead end An acoustical treatment plan for rooms, in which one end is highly absorbent and the other end is reflective and diffusive.

logarithm An exponent of 10 in the common logarithms to the base 10. For example, 10 to the exponent 2 = 100; the log of 100 = 2.

loudness A subjective term for the sensation of the magnitude of sound.

loudspeaker An electroacoustical transducer that changes electrical energy to acoustical energy.

masking The amount (or the process) by which the threshold of audibility for one sound is raised by the presence of another (masking) sound.

mass-air-mass resonance A resonating system composed of the mass of two spaced glass panes, for example, and the air between them. At the frequency at which this system is resonant there is usually a dip in the transmission-loss curve.

mean free path For sound waves in an enclosure, it is the average distance traveled between successive reflections.

microphone An acoustical-electrical transducer by which sound waves in air can be converted to electrical signals.

middle ear The cavity between the eardrum and the cochlea in which the ossicles connect the eardrum and the oval window.

mixer A resistive device, sometimes very elaborate, used for combining signals from many sources.

mode A room resonance. There are axial, tangential, and oblique modes.

monaural See *monophonic*.

monitor Loudspeaker used in the control room of a recording studio.

multitrack A system of recording multiple tracks on magnetic tape or other media. The signals recorded on the various tracks are then "mixed down" to obtain the final recording.

NAB National Association of Broadcasters.

noise Interference of an electrical or acoustical nature. Random noise is a desirable signal used in acoustical measurements.

noise criteria Standard spectrum curves by which a given measured noise may be described by a single NC or NCB number.

nonlinear A device or circuit is nonlinear if a signal passing through it is distorted.

normal mode A room resonance. See *Mode*.

null A low or minimum point on a graph. A minimum pressure region in a room.

oblique mode See *mode*.

octave The distance between two frequencies having a ratio of 2:1.

oscilloscope A cathode-ray type of indicating instrument.

ossicles A linkage of three tiny bones providing the mechanical coupling between the eardrum and the oval window of the cochlea consisting of the hammer, anvil, and stirrup.

out of phase The offset in time of two related signals.

oval window A tiny membranous window on the cochlea to which the footplate of the stirrup ossicle is attached. The sound from the eardrum is transmitted to the fluid of the inner ear through the oval window.

overtone A component of a complex tone having a frequency higher than the fundamental.

panel absorber A panel mounted with an airspace behind vibrates and absorbs sound energy. A frequency higher than the fundamental.

partial One of a group of frequencies, not necessarily harmonically related to the fundamental, which appear in a complex tone. Bells, xylophones, blocks, and other percussion instruments produce harmonically unrelated partials.

passive absorber A sound absorber that dissipates sound energy as heat.

perforated absorber A panel absorber with an air space becomes a Helmholtz absorber if holes perforate the panel.

phase The time relationship between two signals.

phon The unit of loudness level of a tone.

pink noise A noise signal whose spectrum level decreases at a 3 dB-per-octave rate. This gives noise equal energy per octave.

pinna The exterior ear.

pitch A subjective term for the perceived frequency of a tone.

plenum An absorbent-lined cavity through which conditioned air is routed to reduce noise.

polar pattern A graph of the directional characteristics of a microphone or a loudspeaker.

polarity The relative position of the high (+) and the low (−) signal leads in an audio system.

precedence effect For delay time, less than 50 ms, echos are no longer a noise even if the echo is stronger than the primary sound. This is called the precedence (Haas) effect.

primitive root The root of a prime number.

psychoacoustics The study of the interaction of the auditory system and acoustics.

pure tone A tone with no harmonics. All energy is concentrated at a single frequency.

Q-factor Quality-factor. A measure of the losses in a resonance system. The sharper the resonance curve, the higher the Q.

quadratic residue A mathematical operation used in design of diffusors. There are three 15s in 57, with a residue of 12. The residue is the important part.

random noise A noise signal, commonly used in measurements, which has constantly shifting amplitude, phase, and frequency, and a uniform spectral distribution of energy.

ray At higher audio frequencies sound can be considered as rays traveling in straight lines in a direction normal to the wavefront.

reactive absorber A sound absorber, such as the Helmholtz resonator, which involves the effects of mass and compliance as well as resistance.

reactive silencer A silencer in air-conditioning systems that uses reflection losses for its action.

reflection For surfaces large compared to the wavelength of impinging sound, sound is reflected much as light is reflected, with the angle of incidence equal to the angle of reflection.

reflection-phase grating A diffusor of sound energy using the principle of the diffraction grating.

refraction The bending of sound waves traveling through layered media with different sound velocities.

resistance That quality of electrical or acoustical circuits that results in dissipation of energy through heat.

resonance A natural periodicity, or the reinforcement associated with this periodicity.

resonator silencer An air-conditioning silencer employing tuned stubs and their resonating effect for its action.

reverberation The tailing off of sound in an enclosure after the source has stopped. Caused by multiple reflections from boundaries of room.

reverberation time The time required for the sound in an enclosure to decay 60 dB.

ringing High-Q electrical circuits and acoustical devices have a tendency to oscillate (or ring) when excited by a suddenly applied signal.

round window The tiny membrane of the cochlea that opens into the middle ear cavity serving as a pressure release for the cochlea fluid.

RT60 Reverberation time.

sabin The unit of sound absorption. One square foot of open window has an absorption of 1 sabin.

semicircular canals The three sensory organs for balance that are a part of the cochlear structure.

sequence, maximum length A mathematical sequence used in determining the well depth of diffusors.

283

sequence, primitive root A mathematical sequence used in determining the well depth of diffusors.

sequence, quadratic residue A mathematical sequence used in determining the well depth of diffusors.

signal-to-noise ratio The difference between the nominal or maximum operating level and the noise floor in dB.

sine wave A periodic wave related to simple harmonic motion.

slap-back A discrete reflection from a nearby surface.

solid-state The branch of physics having to do with transistors, etc.

sone The unit of measurement for subjective loudness.

sound absorption coefficient The practical unit between 0 and 1 expressing the absorbing efficiency of a material. It is determined experimentally.

sound level meter A microphone-amplifier-meter arrangement calibrated to read sound-pressure level above the reference level of 20 micropascals.

sound power level A power expressed in dB above the standard reference level of 1 picowatt.

sound spectrograph An instrument that displays the time, level, and frequency of a signal.

spectrum The distribution of the energy of a signal with frequency.

spectrum analyzer An instrument for measuring (and usually recording) the spectrum of a signal.

specularity A term devised to express the efficiency of diffraction-grating types of diffusors.

spherical divergence Sound diverges spherically from a point source in free space.

splaying Walls are splayed when they are constructed somewhat "off-square", i.e., a few degrees from the rectilinear form.

standing wave A resonance condition in an enclosed space in which sound waves traveling in one direction interact with those traveling in the opposite direction, resulting in a stable condition.

STC Sound transmission class: a single-number system of designating sound transmission loss of partitions, etc.

steady-state A condition devoid of transient effects.

stereo A stereophonic system of two channels.

superposition Many sound waves may traverse the same point in space, the air molecules responding to the vector sum of the demands of the different waves.

tangential mode A room mode produced by reflection off four of the six surfaces of the room.

threshold of feeling (pain) The sound pressure level that makes the ears tickle, located about 120 dB above the threshold of hearing.

threshold of hearing The lowest sound level that can be perceived by the human auditory system. This is close to the standard reference level of sound pressure, 20 micropascals.

timbre The quality of a sound related to its harmonic structure.

tone burst A short signal used in acoustical measurements to make possible differentiating desired signals from spurious reflections.

tone A tone results in the auditory sensation of pitch.

tone control An electrical circuit to allow adjustment of frequency response.

transducer A device for changing electrical signals to acoustical or vice versa, such as a microphone or loudspeaker.

transient A short-lived aspect of a signal, such as the attack and decay of musical tones.

treble The higher frequencies of the audible spectrum.

volume The colloquial equivalent of sound level.

watt The unit of electrical or acoustical power.

wave A regular variation of an electrical signal or acoustical pressure.

wavelength The distance a sound wave travels in the time it takes to complete one cycle.

weighting Adjustment of sound-level meter response to achieve a desired measurement.

white noise Random noise having uniform distribution of energy with frequency.

References

Allison, Roy F. 1979. Influence of listening room on loudspeaker systems. *Audio.* (63) 8, August 1979. pp. 37–40.

American Society of Heating, Refrigerating, and Air-Conditioning Engineers, Inc. (ASHRAE). 1984. *ASHRAE Handbook—Systems.* Atlanta, GA: ASHRAE, Inc.

ASTM. *Determination of Sound Transmission Class.* American Society for Testing Materials (specification document). Designation E413-70T.

Beranek, L. L. 1962. *Music, Acoustics, and Architecture.* New York, Wiley.

Beranek, L. 1989. Balanced noise-criterion (BNC) curves. *Journal of the Acoustic Society of America.* (86) 2, August 1989. pp. 69–101.

Blazier, W. Jr., and R.B. DuPree. 1994. Investigation of low-frequency footfall noise in wood-frame multifamily building construction. *Journal of the Acoustic Society of America.* (96) 3, September 1994. pp. 1521–1532.

Cops, A., H. Myncke, and G. Vermeir. 1973. Insulation of reverberant sound through double and multilayered glass constructions. *Acustica.* 23, pp. 257–265.

D'Antonio, P. and J.H. Konnert. 1984. The reflection phase grating: design, theory, and application. *Journal of the Audio Engineering Society.* (32) 4, April 1984. pp. 228–238.

D'Antonio, P. and J.H. Konnert. 1992. The QRD Diffractal: a new one-dimensional fractal sound diffusor. *Journal of the Audio Engineering Society.* (40) 3, March 1992. pp. 117–129.

D'Antonio, P. and J.H. Konnert. 1985. Recording control room design incorporating a reflection-free zone and reflection phase grating acoustical diffusors. Paper presented at the 110th meeting of the Acoustical Society of America. *Journal of the Acoustic Society of America* supplement. (1) 78, November 1985. p. 59.

D'Antonio, P., C. Bilello, and D. Davis. Optimizing home listening rooms, Part 1. Presented at the 85th Convention of the Audio Engineering Society, preprint 2735.

D'Antonio, P., J. Konnert, and R. Berger. 1988. Control room design utilizing a reflection-free zone and reflection phase-grating diffusors: a case study. *78th Audio Engineering Society Convention.* Anaheim, CA. May 1988.

Davis, D. and C. Davis. 1980. The LEDE™ concept for control of acoustic and psychoacoustic parameters in recording control rooms. *Journal of the Audio Engineering Society.* (28) 9, September 1980. pp. 585–595.

Egan, M.D. 1972. *Concepts in Architectural Acoustics.* New York, NY: McGraw-Hill, Inc.

Gilford, C. 1972. *Acoustics for Radio and Television Studios.* London, UK: Peter Peregrinus, Ltd.

Grantham, J.B. and T.B. Heebink. 1973. Sound attenuation provided by several wood-frame floor/ceiling assemblies with troweled floor toppings. *Journal of the Acoustic Society of America.* (54) 2, 1973. pp. 353–360.

Green, D.W. and C.W. Sherry. 1982. Sound transmission loss of gypsum wallboard partitions. *Journal of the Acoustic Society of America.* Report #1: Unfilled steel stud partitions. (71) 1, January 1982. pp. 90–96. Report #2: Steel stud partitions having cavities filled with fiber batts. (71) 4, April 1982. pp. 902-907. Report #3: 2 × 4 wood stud partitions. (71) 4, April 1982. pp 908-914.

Harris, C.M. 1957. *Handbook of Noise Control.* New York, NY: McGraw-Hill, Inc.

Hedeen, Robert A. 1980. *A Compendium of Materials for Noise Control,* DHEW (NIOSH) publication No. 80-116, U.S. Government Printing Office, out of print

Hirschorn, M. 1994. Fiberglass & noise control—is it a safe combination? *Sound and Vibration.* (28) 10, October 1994. pp. 6–10.

Jackson, G.M. and H.G. Leventhall. 1972. The acoustics of domestic rooms. *Applied Acoustics.* 5, 1972. pp. 265–277.

Libby-Owens-Ford Company (LOF). *Sound Reduction Design Considerations for Construction Glass.* Toledo, OH: Libby-Owens-Ford Company.

Lockner, J.P.A. and J.F. Burger. 1958. The subjective masking of short time-delayed echoes by their primary sounds and their contribution to the intelligibility of speech. *Acustica.* (8), 1958. pp. 1–10.

Mankovsky, V. S. 1971. *Acoustics of Studios and Auditoria,* Focal Press, Ltd.

Meyer, E. and G.R. Schodder. 1952. *On the influence of reflected sound on directional localization and loudness of speech,* Nachr. Akad. Wiss., Göttingen, Math. Physics, Klasse IIa, 6, 31–42.

Morse, P. and R. H. Bolt. 1944. Sound waves in rooms. *Reviews of Modern Physics.* (16) 2, April 1944. pp. 69–150.

National Bureau of Standards (NBS). 1975. *Acoustical and thermal performance of exterior residential walls, doors, and windows.* NBS Technical Publication PB-246-716, 158 pp.

Northwood, T.D. 1968. *Transmission Loss of Plasterboard Walls.* Building Research Note no. 66. Ottawa, Canada: Division of Building Research, National Research Council.

287

Olive, S.E. and F.R. Toole. 1989. The detection of reflections in typical rooms. *Journal of the Audio Engineering Society.* (37) 7–8, July 1989. pp. 539–553.

Pelton, H.K., S. Wise, and W. Sims. 1994. Active HVAC noise control systems provide acoustical comfort. *Sound and Vibration.* (28) 7, July 1994. pp. 6–13.

Pisha, B. and C. Bilello. 1987. Designing a home listening room. *Audio.* August 1987. pp. 48–58.

Quirt, J.D. 1982. Sound transmission through windows: I. single and double glazing. *Journal of the Acoustic Society of America.* (72) 3, September 1982. pp. 834–844.

Quirt, J.D. 1983. Sound transmission through windows: II. double and triple glazing. *Journal of the Acoustic Society of America.* (74) 2, August 1983. pp. 834–844.

RPG Diffusor Systems, Inc. 1990. *The RPG Home Concert Hall.* RPG Diffusor Systems, Inc. Upper Marlboro, MD.

Ruzicka, J.E. 1971. Fundamental concepts of vibration control. *Sound and Vibration.* (5) 7, July 1971. pp. 16–23.

Sabine, Hale et al. 1975. *Acoustical and thermal performance of exterior residential walls, doors, and windows.* National Bureau of Standards (NBS) Technical Publication PB-246-716.

Sanders, G.J. 1968. Silencers: their design and applications. *Sound and Vibration.* (2) 2, February 1968. pp. 6–13.

Schroeder, M.R. 1975. Diffuse sound reflections by maximum-length sequence. *Journal of the Acoustic Society of America.* (57) 1, January 1975. pp. 149–150.

Schroeder, M.R. 1979. Binaural dissimilarity and optimum ceilings for concert halls: more lateral sound diffusion. *Journal of the Acoustic Society of America.* (65) 4, April 1979. pp. 958–963.

Schroeder, M.R. 1988. *Number Theory in Science and Communication, 2nd. ed.* New York, NY: Springer-Verlag.

Seikman, William. 1969. *Outdoor acoustical treatment,* Journal of the Acoustic Society of America, (46) 4 (Part I).

Sepmeyer, L.W. 1965. Computed frequency and angular distribution of the normal modes of vibration of rectangular rooms. *Journal of the Acoustic Society of America.* (37) 3, March 1965. pp. 413–423.

Volkmann, J. 1941. Polycylindrical diffusers in room design. *Journal of the Acoustic Society of America.* (13), January 1941. pp. 234–243.

Warnock, A. 1983. *How to Reduce Noice Transmission Between Apartments.* Building Research Note no. 44. Ottowa, Canada: Division of Building Research, National Research Council.

Resources

Acoustic Sciences Corporation (ASC)
PO Box 1189
Eugene, OR 97440
voice: (800) 272-8823
fax: (503) 343-9245

Brejtfus
3844 E. University Drive
Suite 2
Phoenix, AZ 85034
voice: (800) 264-9190
fax: (602) 470-8588

Brüel & Kjaer
World Headquarters
Naerumn Denmark
DK 2850
voice: (714) 978-8066

Flat Glass Marketing Association
3910 Harrison St.
Topeka, KS 66601
voice: (913) 266-7013
fax: (913) 266-0272

Illbruck Acoustic Products (makers of SONEX)
3800 Washington Ave. N.
Minneapolis, MN 55412
voice: (800) 662-0032
fax: (612)-552-5639

Larson-Davis, Inc.
1681 West 820 North
Provo, UT 84601
voice: (801) 375-0177
fax: (801) 375-0182
telex: 705560

Libby-Owens-Ford Company (LOF)
811 Madison Ave.
PO Box 799
Toledo, OH 43695
voice: (419) 247-3731

Overly Manufacturing Company
PO Box 70
Greensburg, PA 15601
voice: (412) 834-7300
fax: (510) 468-0539

Owens-Corning
Fiberglas Tower
Toledo, OH 43659

Quest Technologies
510 West Worthington St.
Oconomowoc, WI 53066
voice: (800) 245-0779
fax: (414) 587-4047

R.P.G. Diffusor Systems, Inc.
651-C Commerce Drive
Upper Marlboro, MD 20773
voice: (301) 249-0044
fax: (301) 249-3912

Tectum Inc.
PO Box 920
Newark, OH 43055
voice: (614) 345-9691

Index

Illustrations are indicated in **boldface.**

293

295

About the Author

F. Alton Everest has earned a place as acoustical expert to the field of audio, sound recording, and high fidelity through his books and articles. From his *Acoustical Techniques for Home and Studio* (1973) through 3 editions of *Master Handbook of Acoustics* (latest 1994), he has led sound engineers and amateurs through the acoustical thickets of sound recording and studio design.

He has the scholastic underpinnings (BSc in EE, Oregon State; EE, Stanford) and the professional experience (Professor, Oregon State and Hong Kong Baptist Universities, 25 years film production, 16 years consulting in acoustics, and 4 years in wartime research in undersea acoustics), combined with the drive to lend a published hand to those in audio seeking help in understanding the seemingly abstract field of acoustics.

F. Alton Everest is an Emeritus Member of the Acoustical Society, A Life Member of the IEEE, a Life Fellow of Society of Motion Picture and Television Engineers, Member of Audio Engineering Society, and Co-Founder and Past President of American Scientific Affiliation.